No. 3110
$19.95

The Complete
Electronics
Career Guide

Joe Risse

TAB **TAB BOOKS Inc.**

Blue Ridge Summit, PA

FIRST EDITION
FIRST PRINTING

Copyright © 1989 by TAB BOOKS Inc.

Library of Congress Cataloging-in-Publication Data

Risse, Joseph A.
    The complete electronics career guide / by Joe Risse
      p. cm.
Includes index.
ISBN 0-8306-1410-9   ISBN 0-8306-3110-0 (pbk.)
    1. Electronics—Vocational guidance. I. Title.
  TK7845.R57 1989
621.381′024—dc19   88-31770
                            CIP

TAB BOOKS Inc. offers software for sale. For information and a catalog, please contact
TAB Software Department, Blue Ridge Summit, PA 17294-0850.

Questions regarding the content of this book should be addressed to:
TAB BOOKS Inc.
Blue Ridge Summit, PA 17294-0214

Edited by Alyson Grupp

Cover photograph courtesy of TEKTRONIX, INC.

Acknowledgments

W RITING A BOOK IS NOT POSSIBLE WITHOUT COOPERATION OF other people and organizations. Some of the people and organizations that provided me with special assistance are: Dan Kazmierski, Marlene Noteware, and Rebecca Warwick of ICS-Intext; Dick and Dottie Glass, and Ron Crow of ETA: Barbara Rubin of ISCET; Nancy Rodgers of A P Products; the Electronic Industries Association; AMPEX, Inc.; Kevin Toolan of the Tobyhanna U.S. Army Depot; Donna Sedor of C-TEC Corp.; Sidney Bates of the Radio Shack division of Tandy Corporation; Wally S. Harrison, editor of NESDA and ISCET *Professional Electronics* magazine; the National Aeronautics and Space Administration; Janelle Rice of Hewlett-Packard Company; John Porter of NAP Consumer Electronics Corp.; Paul Nies, L. Bowden, and Chuck Robertson of SENCORE; Debra Seifert and Dani Rhodes of TEKTRONIX, INC.; Larry Kay and Dan Wright of the John Fluke Mfg. Co; Bernard Surtz of Andrew Corporation; Laird Campbell, editor, QST, ARRL; good author friends J.A. "Sam" Wilson, and Edward Noll; RCA Astroelectronics; Art Williams of Johnson Technical Institute; Sharon Cannon of the Society of Broadcast Engineers; my son Bill and his wife Karen for valuable graphic assistance; Dr. Eugene McGinnis, University of Scranton; Leonard Valore, National Education Corporation; and VGM Career Horizons, National Textbook Company.

*Dedicated to
my wife, Anne Stegner Risse,
and our grandsons
Matthew Joseph Risse and Stephen Vincent Armitage*

Contents

Preface

I T IS MY INTENTION IN THIS BOOK TO ANSWER AS MANY QUESTIONS AS possible from those considering electronics as a career, and those already working in electronics. Questions from production workers, engineers, technicians, technical writers, parts and sales personnel, and others have been considered. My years of experience in electronics service and engineering and in electronics education have given me insights about what electronics workers and prospective workers want to know.

Since much of the information provided here is sometimes hard to find, electronics educators, suppliers, parts and counter persons, guidance personnel, instructors, professors, and school, public, and company librarians might want to keep this book handy as a reference.

Introduction

T HIS BOOK PROVIDES INFORMATION ABOUT THE FIELD OF ELEC-
tronics for those who are considering a career in electronics
—as electronics mechanics, technicians, technologists, or
engineers—as well as for those who are already working in that
field and wish to further their careers.

Chapter 1 tells what the field of electronics is all about. It iden-
tifies several jobs and positions, and introduces ways to get education and training in the electronics field. The chapter also tells
about some applications of electronics.

Chapter 2 describes how electronics hobbies like amateur
radio, kit building and experimentation, citizens' band radio, and
use of personal computers help enhance or begin a career in
electronics.

Chapter 3 describes employment in several areas of electronics
including industry, broadcasting, communications, cable television, and computers and data processing.

Chapter 4 discusses positions available in the electronics industry for engineers, technicians, and assemblers.

Chapter 5 discusses troubleshooting jobs available for technicians and engineers.

Chapter 6 describes work opportunities in broadcasting and
cable TV for engineers and technicians.

Chapter 7 describes consumer electronics, including entertainment electronics, and discusses the wide range of opportunities that exist in that field.

Chapter 8 describes positions available in telephony, which now overlaps into data communications, two-way radio, and cellular radio.

Chapter 9 offers information on opportunities in consumer servicing and troubleshooting.

Chapter 10 describes the rewards available to those who combine a good knowledge of electronics with an ability to write about electronics, or to teach others about the field.

Chapter 11 expands on what was introduced earlier on the types and levels of education and training available for work in the electronics field. Chapter 12 concentrates on ways of advancing in a job or career in the electronics field. Chapter 13 lists the various types of technical literature available, including magazines, catalogs, brochures, and pamphlets, and describes how to get it. Chapter 14 tells how you can enrich your career by becoming a member of a technical association or society serving those working in electronics as engineers or technicians.

Chapter 15 tells of the advantages and risks of advancing your career in electronics through job hopping (changing jobs and employers on a frequent basis). Some tips are provided on finding a job, including information on developing a resumé covering your education, training, experience and personal qualifications. A sample resumé and a sample cover letter are included.

The appendices contain names and addresses of publications, manufacturers, and associations, as well as other educational resources.

❖ 1

The Electronics Industry

THE PURPOSE OF THIS BOOK IS TO TELL YOU ABOUT THE CAREERS AND jobs available in the field of electronics. Electronics is an extension of electricity and its general principles. The same electricity that comes from the wall receptacle in your home and from the battery in your automobile has many applications; electronics is one of them.

Electronics makes radio and television transmission and reception possible. Electronic equipment in a TV studio is shown in FIG. 1-1. Electronics can control manufacturing and industrial processes, and electronic integrated circuits (ICs), like the IC in FIG. 1-2, have made typewriter-size computers possible. One IC can contain up to thousands of individual circuits.

The term *electronics* comes from the word "electron." An electron is a very tiny and invisible electrical quantity which is part of an atom. Electronics involves the control of electrons. The main devices in which electrons are controlled are vacuum tubes, transistors, integrated circuits, and other semiconductor or solid-state devices.

HISTORY OF ELECTRONICS

The modern age of electronics started with the early days of radio. Although the electron was not discovered until 1897 in England by Sir Joseph John Thompson, Thomas Edison controlled electrons for the first time in 1883 by causing them to jump across a partial vacuum between the filament and a metal plate inside a light bulb. In 1894 *electronic communications*, the transfer of information over electrical circuits, was achieved by Oliver Lodge in England with a system of wireless communications. Ernest Rutherford, Guglielmo Marconi, and others made further progress in wireless telegraphy around 1900. Then in 1906, Dr. Lee De Forest invented the three-element vacuum tube, which made possible the development of radio broadcasting and later television. (Vacuum tubes are rarely found in present electronic equipment, having been replaced by transistors and integrated circuits.)

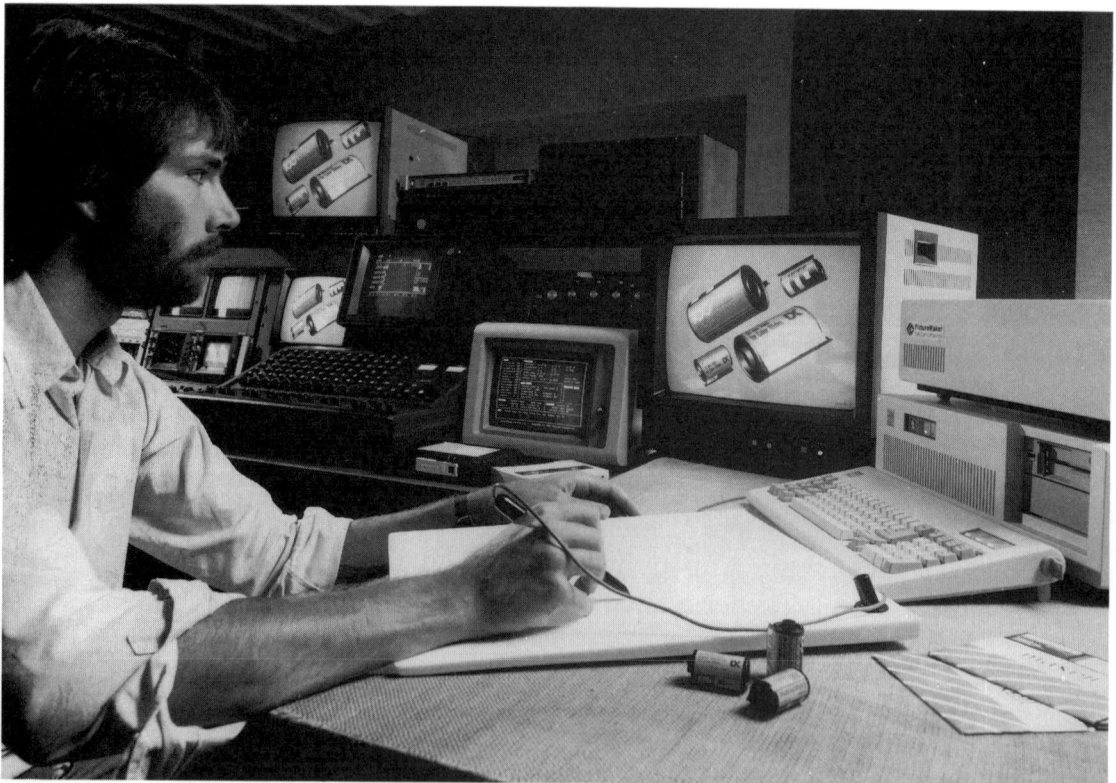

Fig. 1-1. The equipment shown here produces special effects in TV programs. COURTESY AMPEX CORPORATION AND CUBICOMP CORPORATION.

When the *Titanic* sank in 1912, wireless communications saved many lives. Wireless communications was also used by the United States and others during World War I. Shortly thereafter, radio broadcast stations were in operation throughout the United States. Regularly scheduled programs by radio station KDKA in Pittsburgh, Pennsylvania, in the standard amplitude modulation (AM) radio band, started in 1920. Commercial frequency modulation (FM) radio stations started in 1939, and stereo FM broadcasting began in 1961. Commercial television broadcasting started in 1941 but did not make substantial progress until after World War II.

The popularity of television receivers dates from around 1945. Color television as we know it today started in 1953, although there was some limited use earlier of other color TV systems.

Rapid advancement in electronic applications occurred during World War II in the form of equipment for guidance, surveillance, detection, and communications. Radar and sonar were used to navigate in air and on sea and to locate objects and targets. Other electronic equipment was used for weapons fire control, giant calculators, navigation, and automatic flight control.

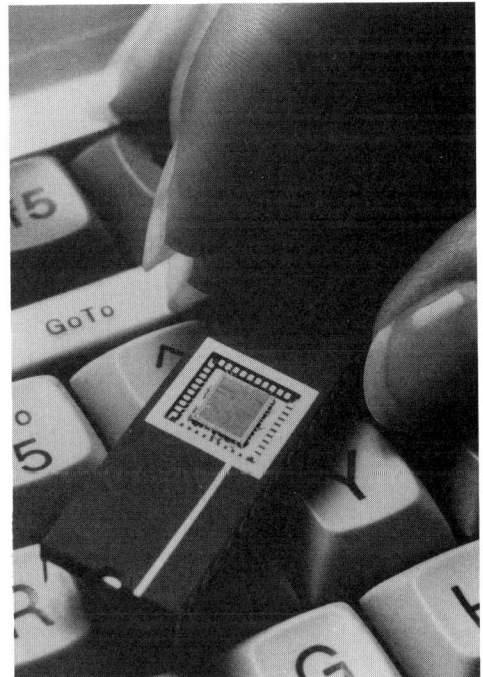

Fig. 1-2. Some integrated circuits are even smaller than this one, being held in a person's fingers. Before integrated circuits, equivalent electronic circuits could occupy several cubic feet. COURTESY OF HARRIS CORPORATION.

Following World War II, FM and TV broadcasting grew rapidly, and in 1947 the development of the transistor signalled the start of the miniaturization of electronic equipment. (A transistor is much smaller and lighter than a vacuum tube.) The transistor and other semiconductor devices soon resulted in hand-held, battery-operated radio receivers and smaller computers. Soon, ways were developed to fabricate transistors, capacitors, resistors, and other components from tiny chips of silicon so that many circuit devices were concentrated into a single device, called an *integrated circuit* (IC).

The development of the transistor and, later, the IC led to the rapid growth of smaller and cheaper consumer, industrial, and commercial electronic equipment and systems. In particular, the IC spurred the growth of computers small enough to be used on desktops and in homes, home-satellite systems, TV remote controls, personal calculators, digital watches, and medical instruments. Modern video cassette recorders (VCRs) are the outgrowth of video tape recorders, which had their practical beginning in 1956 (FIG. 1-3).

The applications and uses of electronics seem endless, and electronics can offer you a promising future — if you prepare for it.

OPPORTUNITIES IN ELECTRONICS

Jobs and positions available in electronics range from those that require little or no background and training, to those that demand advanced college training at the highest level. For every engineer and scientist, several technicians are usually required. Craftsmen and plant engineers are also needed; for example, to carry out preventive and corrective routine maintenance in industry (FIG. 1-4) and to install electronic systems. These craftsmen may or may not be members of a union.

A great many electronic technicians are employed in the service and repair of consumer electronic equipment such as radio and TV sets, VCRs, CB radios, stereos, calculators, computers, video games, watches, security systems, electronic health-care products, etc. Other people are employed by parts distributors and suppliers to buy and sell parts for these products.

One fast-growing area of electronics is the installation, operation, and servicing of earth stations that transmit and receive TV, data, telephone, and other signals to and from satellites orbiting the earth. Many of these satellite systems are owned or operated by radio and TV stations, communications companies, movie services, cable TV operations, and data distribution concerns. Home-satellite earth stations are sometimes referred to as *Television Receive-Only satellite earth stations (TVROs)*. TV satellites are orbited by the space project (FIG. 1-5). Proper installation and performance of earth satellite systems requires the service of experienced and skilled technicians and engineers.

Fig. 1-3. The invention of the video tape recorder led to many new jobs in electronics. The "announcement event" by videotape pioneer, Ampex, is shown here. It happened in Chicago in 1956. COURTESY OF AMPEX CORPORATION.

In the field of telecommunications, telephone companies employ a considerable number of workers. Most switching systems are now electronically operated. Push-button dialing on ordinary telephones requires electronic equipment in both the telephone itself, as well as at central offices and switching centers. Cordless or wireless telephones are really radio transmitter and receiver instruments which, of course, are electronic devices. There are also telephone answering machines, amplified telephones, and videophones (which provide pictures as well as sound).

Fig. 1-4. Virtually all industrial equipment uses electric and electronic controls which are kept running by plant engineers who have a strong electronic background. COURTESY OF JOHN FLUKE MFG. CO.

Opportunities exist for work on or with communications equipment such as CB radios, cellular radios, mobile radio-telephones, and two-way radio systems used by police, fire, and public safety organizations. Utilities and service organizations are also heavy users of two-way radio equipment.

Other areas of communications employ people who have skill and training in electronics. Broadcasting is one example. Radio and TV broadcast equipment, especially TV equipment produced since the 1950s (FIG. 1-6) is quite complex. For maintenance and

Fig. 1-5. Many technicians and engineers work in the space industry, which includes designing, building, launching, and utilizing communications satellites such as this one shown leaving the bay of the shuttle-orbiter, *Discovery*. COURTESY OF RCA ASTROELECTRONICS.

7

troubleshooting, broadcasters hire highly skilled people to keep their equipment operating properly and to repair it quickly when needed. At one time, an FCC license was needed to work on radio and TV transmitters, but a government license is no longer needed. Instead, owners of broadcast stations now have the full burden of determining that the people they employ are competent, and able to adjust, maintain, and repair their transmitter and antenna systems. Station operators often try to hire technicians and engineers who have been certified by the Society of Broadcast Engineers in order to assure themselves that their equipment will be in the hands of properly qualified people.

Facsimile and electronic mail are two other areas of electronic communications. These services provide for the transmission of printed and written documents or other graphic matter. Many corporations use these systems for the rapid transmission of in-

Fig. 1-6. TV broadcast engineer making the first use of a practical video-tape recorder, on November 30, 1956. Previously, TV shows were recorded on film in order to be saved, which usually did not result in the best quality. COURTESY OF AMPEX CORPORATION.

formation from one point to a distant point; for example, pages of accounting data, a corrected page of a manuscript, a copy of a diagram or drawing, a schematic, or an urgently needed report.

Advancements in the development of fiberoptics, (optical fibers that can transmit light over long distances) have opened up another area of electronic communications. In many cases, fiber-optical equipment is being used to replace systems that in the past used copper wires for information transmission. *Optical fibers* are plastic threads that are smaller and lighter than copper wires, and they can carry more channels of information for the same diameter of conductor. (Fiberoptics offers many advantages in noncommunications applications, also. For example, doctors can see inside the body or concentrate a high-intensity laser beam for surgical use.)

Opportunities in industrial electronics are also many and varied. Industrial electronics includes computer and data-processing equipment, testing and measuring apparatus, nuclear electronic devices, medical equipment, programmable controls, robotic control and automation equipment, and control and processing equipment. Industrial electronic equipment of various types currently costs more than $68 billion, and increases of about 15 percent each year are typical. Computer equipment accounts for about $45 billion of this amount.

Electronics has brought about the automation of business offices, along with career opportunities for sales and service engineers and technicians to design, sell, install, service, adjust, and repair these systems. Electronic office equipment includes information entry and processing equipment, communications systems, and records management equipment. Examples of information entry and processing equipment are electric and electronic typewriters, dictation equipment, word processors (FIG. 1-7), and small business computers. Communications equipment includes private branch exchange (PBX) telephone equipment, copiers, telephone dialing systems, facsimile and electronic mail machines, answering machines, key systems, and teleconferencing systems. Records management equipment includes data display and storage equipment, computerized inventories, and other electronic libraries.

A rapidly growing area of electronics involves the equipment used by banks. This rapid growth is partly a result of the trend toward electronic fund transfers (EFT). EFT equipment includes automated teller machines, cash dispensers, point-of-sale (POS) terminals, automated clearing houses, and automated payment of telephone and other utility bills. Most federal government payments are now made electronically. Automated teller machines are used by banks, credit unions, and savings and loan associations.

Fig. 1-7. Computer operators are workers in the field of electronics. COURTESY OF RADIO SHACK, A DIVISION OF TANDY CORPORATION.

Career opportunities also exist with computer-aided publishing (CAP) equipment used by publishers. Using CAP, editors and graphic designers can work with systems that automate selection of type fonts, graphics, layout, and photographs (halftones).

Since the early 1970s, manufacturers have been using industrial robots in a number of manufacturing jobs. Robots have been used on a large scale in foreign countries for some time, especially in Japan. Now their use is growing fast in the United States, especially in the automotive industry. An experimental robot is shown in FIG. 1-8. Designing, installing, maintaining, and troubleshooting these systems provides unusually interesting careers in electronics.

Bar codes have been used for several years now in many grocery checkout counters. Bar codes activate electronic circuits which indicate price, name of item, inventory and reorder status, and other information. They are also used by industry and the military to automate parts and materials tracking. The bars and

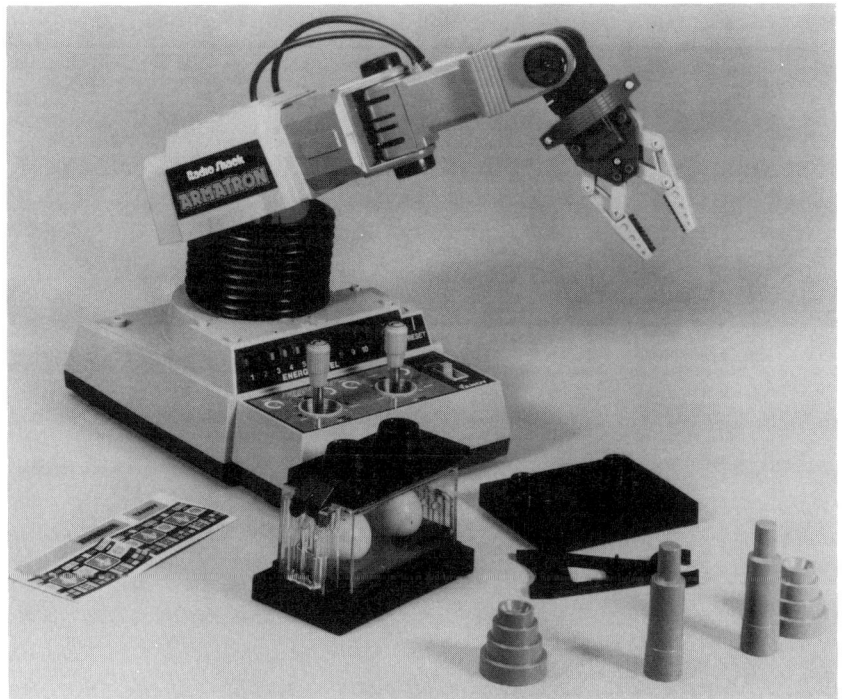

Fig. 1-8. Experimental robot for students, educators, hobbyists, or designers. COURTESY OF RADIO SHACK, A DIVISION OF TANDY CORPORATION.

spaces of a bar code are translated by a computer with special scanning devices into digital data that indicate whatever data is needed in the operation.

Bar codes are also used by airlines to sort baggage, by blood banks to classify blood, by libraries and hospitals, by buyers for item selection in sales catalogs, by cafeterias to keep track of people, etc.

The use of computers in computer-aided design (CAD), computer-aided manufacturing (CAM), and computer-aided engineering (CAE) systems has greatly increased the need for engineers and technicians to install, operate, and troubleshoot these systems. Programmers and system planners are also needed.

GETTING STARTED IN ELECTRONICS

For most types of work in electronics, it is important to develop a skill and interest in mathematics, in middle school grades and continuing through high school. However, even if you didn't have a real interest or ability in basic mathematics in your earlier school years, it's not too late. Remedial mathematics classes are available through either home study or adult evening classes in local high schools, vocational-technical schools, colleges, and junior colleges.

In your educational plans, it is helpful to include courses in science, physics, and English. Science and physics courses are important because electronics is based on scientific and physical principles. The study of English is important so you can get the most out of reading and studying about electronics, but also so you can get your ideas across to others. This transfer of ideas is often essential on the job; for example while discussing your work with your employer, or when making up a report.

The next question is, how do you actually obtain skills in electronics, in addition to the educational prerequisites? A child can start out early to obtain exposure to electronics by building electronics kits and experimenting outfits available at toy, hobby, and electronics parts suppliers such as Radio Shack and others.

As a student gets a little older, more advanced kits are available from the same sources. If you're sure that you do not plan to attend college later, you can consider selecting a vocational high school that offers an electronics program. Some of these high schools provide you with sufficient knowledge and skill to enter the job market as a craftsman. Of course, some graduates of vo-tech schools do go on to college after graduation, especially if they have also concentrated on mathematics and sciences during high school or have taken brush-up and make-up courses.

Other ways to get started and obtain education and experience in electronics include attending trade schools, technical institutes, junior colleges, and colleges. These schools can prepare you for a career as a technician or engineer in the electronics field.

If your plans do not include education beyond high school, at least not right away, you can obtain hands-on experience in electronics through building the kinds of kits mentioned earlier. Educational kits and self-study courses are also available from the Heath Company (Benton Harbor, Michigan) and similar suppliers. Experience and knowledge gained through kit building and experimenting can be listed on job applications. They are sometimes sufficient to get a foot in the door when applying for your first job in electronics.

Many people who are not college-bound obtain knowledge and experience in electronics by joining a local amateur radio club. If no club is available locally, you can still obtain information on amateur radio from publications by the American Radio Relay League (ARRL) (Newington, Connecticut, 06111). Amateur radio operators often build their own transmitter equipment or some of the peripheral items that are part of an amateur station. As you advance in skill and ability, you can earn higher grades of operating licenses, which can also be used as evidence of experience when applying for a job or position in the electronics field.

Enrolling in a home-study course (FIG. 1-9) is another way of obtaining knowledge and some experience in electronics. Home-study courses, also called correspondence or independent-study courses, can provide a worthwhile stepping-stone toward work in electronics, especially if you are seriously dedicated to completing the course. It's important to establish a regular study schedule and have a place to study without undue interruptions. Also, in order to carry out the electronic experiments and projects, you'll need a dry tabletop or workbench with a couple of nearby electrical outlets. To obtain the names of accredited correspondence schools, contact the National Home Study Council (see in Appendix A).

A handbook of accredited residential trade and technical schools is available on request from National Association of Trade and Technical Schools (NATTS) (see Appendix A). You can obtain information on the technical institutes, junior colleges, and colleges in your state or in other states from the department of education in your state's capitol.

ELECTRONICS APPLICATIONS

The electronics industry consists, not only of equipment and component manufacturers, but also distributors, manufacturers' representative firms, consultants, service organizations, and

Fig. 1-9. Wearing a watch or ring is not permitted for technicians in some shops. The danger of shock or severe burn becomes more serious when jewelry is worn. COURTESY OF NESDA/ISCET.

users of electronic products (consumer, industrial, commercial, scientific, government, and military users. Electronic technicians, craftsmen, and engineers are employed in most parts of the electronics industry (FIGS. 1-10 and 1-11).

There is hardly an area of our daily activities that is not touched by electronics. For example, the land transportation field uses electronics extensively. Trains, trucks, buses, and taxicabs use two-way radio for contact with terminals and with each other. The Metroliners operating between New York and Washington depend heavily upon electronics for timing and switching. Also, railroads use microwave radio relay systems, rather than wireline cable and telephone circuits.

The maritime industry handles ship-to-ship and shore-to-shore communications by radio. They also use radar and long-range navigation (loran) for navigation, electronic depth sounders, and ship intercommunications.

Fig. 1-10. Technician troubleshooting computer equipment with multi-trace oscilloscope. (COURTESY OF TEKTRONIX, INC.).

The aviation and aerospace industries use electronics extensively both on the ground and in the air. Aside from control-tower operations and communications with incoming and outgoing aircraft, passenger terminals use electronic computers for speedy designation of passenger seating assignments on both originating flights and subsequent connecting flights right through to the destination. The aerospace industry includes not only airlines but also general aviation airports, airframe manufacturers, satellites, and ground facilities. The space industry could not exist without electronics; neither could the satellite communications industry. Orbiting communications satellites now provide most long-distance telephone and other communications by electronic methods.

In the petroleum industry, electronics controls the flow of different grades of petroleum, and is used by offshore drilling operations.

Federal, state, and local governments expend much of their budgets on two-way radio and other electronic equipment. Some of this equipment includes surveillance equipment, computer

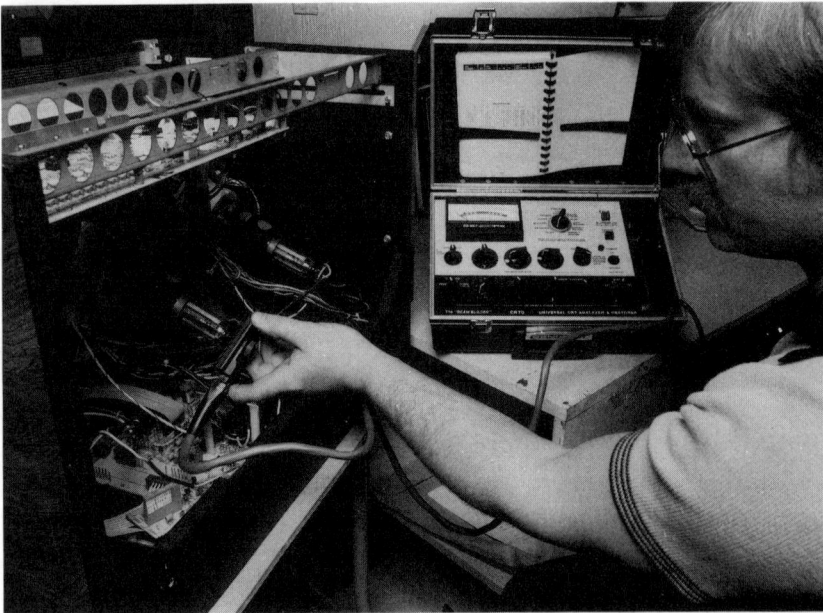

Fig. 1-11. Technicians sometimes must work on high voltage equipment such as this large-screen projection color TV set. COURTESY OF SENCORE ELECTRONIC TEST INSTRUMENTS.

storage of tax and license information, radar speed meters, computers, and traffic signals. Many auto traffic signals are now computer-controlled to enhance the smooth flow of traffic through intersections.

In the medical field, electronic applications and developments are responsible for substantially extending human lifespans. Lasers, fiberoptics, and computers now permit doctors to observe locations inside the human body that were previously accessible only by surgery. Magnetic Resonance Imaging (MRI) equipment allows three-dimensional observation and examination of the brain and other areas inside the body in greater detail and accuracy than is possible with X-ray equipment without the use of radiation.

Hobbies in Electronics

MANY PEOPLE GOT THEIR START IN ELECTRONICS BY BEING EXPOSED to electronics as a hobby, and many of the people who work in electronics have continued their interest in hobby electronics. (Of course, there are many persons not working in electronics who also enjoy electronics as a hobby.)

KIT BUILDING

People who are involved in hobby electronics include builders of predesigned kits that perform some function when completed. Others build circuits that might either be useful, or demonstrate certain electronic principles (FIG. 2-1). Electronics hobbyists include amateur radio operators, or "hams" (FIG. 2-3).

To get information on electronics kits you can build, either for experience or as a hobby, look in your telephone book for the address of the nearest Radio Shack store. Radio Shack can provide you with a catalog and put your name on a mailing list to receive new catalogs and sales announcements as they occur.

Also, check with local electronics parts distributors for predesigned kits for hobbyists and experimenters. Even if the distribu-

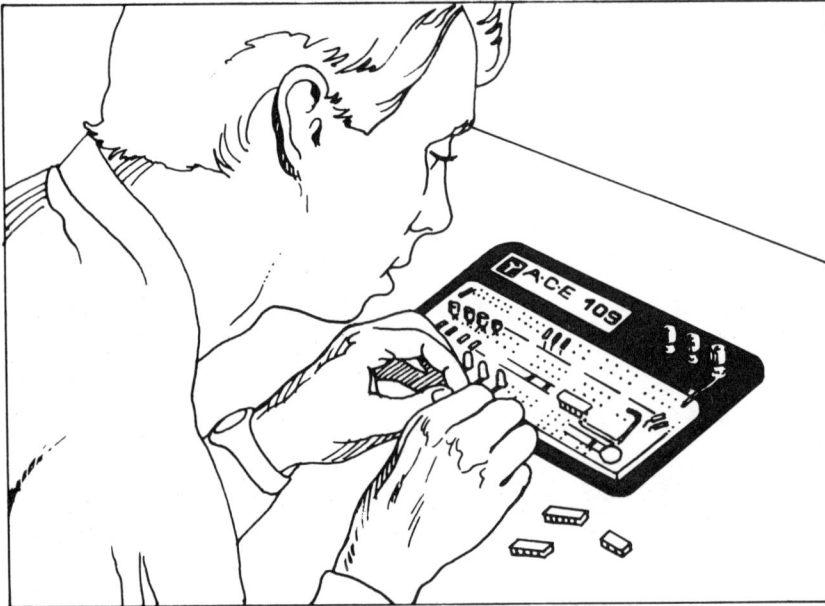

Fig. 2-1. Experimenter shown inserting components into connection block. COURTESY OF A P PRODUCTS, INC.

tors you visit do not have such kits, they can be a convenient source of parts or components in the future. Heath is another company that provides experimental and hobby kits. (See Appendix D). Heath has kits and electronics outfits especially designed for beginning students, as well as for technicians and engineers who want to brush up or catch up on topics they did not learn earlier.

Graymark is another company that will send you a catalog of relatively inexpensive kits that can be assembled and be something useful when finished, such as a burglar alarm, siren, timer, digital car clock, radio receiver, etc. The company also has kits to demonstrate robotics principles. (See Appendix D.)

Other sources of kits and parts and experimenting equipment can be found in electronics magazines. These magazines are discussed in Chapter 13, *Sources of Technical Literature*, and some of them are listed in Appendix B. The magazines themselves often publish articles describing circuits that can be built by experimenters and students. Some additional examples of sources of educational or hobby kits or circuits for building them are in Appendix A.

FIGURE 2-2 shows a connection block with the components for an experiment already plugged in. The components can be removed and other parts inserted for a circuit change.

AMATEUR RADIO

As a radio amateur you will have many interesting and educational activities available to you. You will be able to build, assemble, or install your own equipment; to participate in experiments;

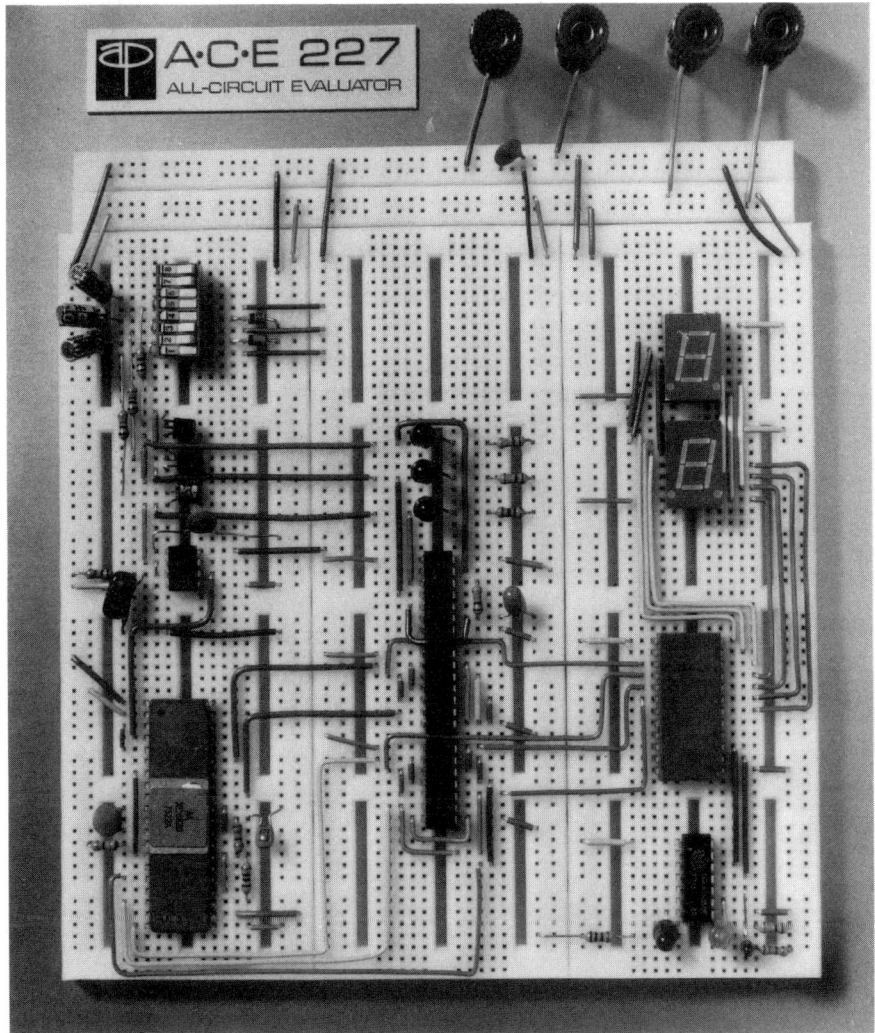

Fig. 2-2. Example of another connection block, with circuit components plugged in. COURTESY OF ASSOCIATED ELECTRONICS/3M A P PRODUCTS: BREADBOARDS, TEST CLIPS, AND ACCESSORIES.

to communicate with other radio amateurs located throughout the world or even in space, by voice and by Morse code; and to communicate with others by television and teletype.

Some amateur radio operators perform emergency and live-saving activities during floods, earthquakes, volcanos, and accidents and illnesses in remote locations by providing communications between rescue, medical, and health authorities.

These are some of the reasons that many people become interested in amateur radio, or why they continue to be active radio operators once they have started. An additional reason, especially for those planning to make electronics a career, is that amateur radio provides training, education, and hands-on experience in electronics.

HOW TO BECOME A RADIO AMATEUR

Amateur radio operation in the United States is governed by the Federal Communications Commission (FCC). The section of the FCC Rules and Regulations dealing with amateur radio is

Fig. 2-3. Many people working in electronics are also active amateur radio operators, or "hams." COURTESY OF ARRL.

known as Part 97. There are five grades, or levels, of amateur radio licenses. You progress from one grade to the next by successfully completing exams of increasing difficulty. Each higher grade license allows the operator greater operating privileges. The license levels (starting with the lowest) are Novice, Technician, General, Advanced, and Amateur Extra.

To obtain the Novice license, you must successfully complete sections of the exam on regulations, on-the-air operating skills, and radio theory. The regulations are based on common sense and are quite easy to learn. The operating skills and the radio theory are also fairly elementary and easy to learn. You'll also be required to pass a Morse code test consisting of sending and receiving letters of the alphabet, the ten numerals, and punctuation characters, at the rate of 5 words per minute or faster. According to the ARRl and individuals who have done it, it is quite easy to learn the code in two weeks or less if you practice regularly. After you've earned your license and have obtained a minimum of equipment, you'll be able to communicate throughout the world with other radio operators using International Morse Code.

There is no charge for taking the license test. If you fail, you can try again after 30 days. More information on taking the Novice exam is available from the ARRL. The ARRL will provide you with the name of a Registered Amateur Radio Instructor in your area who can give you still further information and arrange for you to take the exam when you are ready. The registered instructor can help with your training by providing personalized training and instruction.

If you'd prefer to study on your own, you can obtain a publication, *Tune in the World with Ham Radio*, from ARRL. This publication covers all you need to know to pass the Novice examination. Included with the publications is a Morse code tape that teaches you every letter and numeral, and required punctuation marks.

Another way to become an amateur radio operator is to join a local Amateur Radio club. The ARRL can provide you with a list of 1800 or more such clubs that conduct ten-week classes for beginners. If you learn more quickly, you can take your exam any time you're ready. You'll find that hams are friendly people who speak the language of electronics and who are anxious to help you progress.

The ARRL has available a number of excellent publications and magazines that are authoritative and helpful to radio amateurs and as general references for others in electronics. The ARRL's address and other sources of amateur radio publications can be found in Appendix B.

COMPUTERS

It is possible to acquire an interest and some hands-on experience in electronics if you have the use of a personal computer. Some computers are available now at less than $100. Through the use of such a computer you can become familiar with certain computer terms and expressions that are also an important part of technical electronics. (If you have no practical experience in electronics, it's not a good idea to take a computer out of its case and touch the parts because static electricity from your body could puncture and destroy one or more ICs.)

To become acquainted with the world of personal, or home, computers, you might consider subscribing to one or more of the computer magazines listed in Appendix B. To get an idea of what's in them, before subscribing, you might want to examine one. Most are available at a magazine rack in a local drugstore or in your library.

Aside from magazines on computers, there are also sources that publish other information such as newsletters, catalogs, parts package offers, special equipment listings, and so on. Some of them are listed in the Appendixes; others can be found through the back pages of computer and electronics magazines.

CB RADIO

Many people have gotten into electronics as a career through Citizens Band (CB) radio. To operate a CB radio doesn't require knowledge or skill in electronics, aside from knowing the FCC rules affecting CB radio, but many CB radio hobbyists become intrigued with their equipment and want to know more about their transmitters, receivers, and antennas. They can acquire this knowledge from CB magazines or amateur radio publications and then progress to more technical books or publications. Books on CB radio are available from TAB BOOKS, the publisher of this book.

There are probably CB radio clubs in your area. You can locate these clubs by contacting local dealers and distributors selling CB radios and other electronics equipment.

SOURCES OF PARTS

Parts for electronics hobby, experimenting, and construction projects and for any other purpose can be obtained from sources already mentioned such as Radio Shack stores, local electronics distributors, hobby shops, and (sometimes) radio-TV service shops. Radio-TV shops often throw out TVs and radios that their customers have decided are not worth the cost of having fixed. Some old TVs and radios contain many parts that are still usable.

Parts are also available from mail-order companies. Some of these are listed in Appendix D, *Electronic Parts and Equipment*. You can find other such companies in the classified sections of *Radio Electronics* and other magazines. Some of them offer a grab bag of a thousand or more assorted parts for a very low price.

❖ 3

Employment
in Electronics

YOU ARE ARE PROBABLY READING THIS BOOK BECAUSE YOU ARE
starting to plan or seriously consider a career in the electronics field. You may be thinking about designing, installing, servicing, operating, or selling electronics products. Or you may feel attracted toward teaching electronics or writing about it.

You can start building toward your career immediately if you can decide now what your specific goal is. If you haven't decided, this book may help you to narrow down your choices. However, many people don't really decide on just which area of electronics they want to work in until the opportunity for an attractive position comes along. Often, this is after completion of a course of study in electronics. Opportunities for a good job, or job advancement, or a new and better position, can come along at any time. You help to create these opportunities or to be better able to notice when they exist by being well prepared as far as knowledge, skill, and ability are concerned. Also, you can remain alert to opportunities by knowing where they sometimes exist.

27

There are several major areas in which most of the workers in electronics are employed. These include industry, such as in manufacturing and processing; radio and television broadcasting, including cable television; communications, including two-way radio and telephony; computers and data processing; various government activities; the military; and consumer electronics (sometimes referred to as the home entertainment field). The next few pages introduce some of these areas. Then, later, more detailed information is given.

INDUSTRY

The industrial sections of the U.S. economy, and the economies of other industrialized nations, employ large numbers of electronics engineers (Fig. 3-1), technicians, repairers, and other electronics specialists. Industries that come within this classification include manufacturing industries, such as radio-TV, automobile, and major appliance manufacturers; processing industries, such as petrochemical companies and metal and plastics processing companies; power and energy industries; medical and transportation industries; and many others.

BROADCASTING AND COMMUNICATIONS

Electronics workers are employed by radio and television broadcast stations and cable television systems to install, adjust, repair, and maintain electronic equipment, including TV cameras, TV switchers and monitors, TV receivers, microphones and audio amplifiers, microwave equipment, and so on. The technician in Fig. 3-2 is checking a TV transmission system. Cable TV systems also employ cable installers who install the cables to wire a town or city to distribute the TV programs. They also install the cables connecting from the main cables or trunks to subscribers' homes, offices, or other locations.

Communications electronics includes any applications of electronics that involves the transmission of information through electronic impulses by telephone, telegraph, cable, satellite, radio, or television. These applications include the transmission of funds by banks, electronic mail, telephone systems, remote utility meters/reading, transmission of medical data between hospitals, CB radio, industrial radio, and cellular radio.

Some communications technicians and engineers specialize in installation and adjustment of microwave and satellite dishes, as shown in Fig. 3-3. The dish is being hoisted into position by a helicopter.

COMPUTERS AND DATA PROCESSING

The computer and data processing field is now a major source of employment for electronics technicians, repairers, and engineers. Computers play a vital role in our lives. They are involved in making phone calls, and in processing our paychecks and social security checks. They make and keep track of reservations for

Fig. 3-1. This could be you, working as a quality control or production engineer or technician, completing your report for the day. COURTESY NESDA/ ISCET.

Fig. 3-2. Technician checking out broadcast TV transmission system. COURTESY OF TEKTRONIX, INC.

travel, entertainment, and lodging; they control manufacturing and processing. In many cases, they are used in our homes to keep a record of household expenses, and personal data such as birthdays and doctors' appointments.

Computers come in various sizes ranging from those that can be held on your lap to those that nearly fill a building. Engineers, technicians, and repairers are needed to design, select, install, maintain, use, and repair computers.

Data processing equipment often comes under the classification of computers and also requires workers who can design, select, install, maintain, use, and repair it.

CONSUMER ELECTRONICS

The consumer area represents a huge portion of the electronics industry. Consumer electronics is sometimes referred to as "home entertainment" electronics. However, there are many electronic items used by consumers that are not in the entertainment category; for example, home security systems, cellular telephones, electronic furnace thermostats, and home telephone answering machines.

The manufacture and repair of radios and television still make up a large portion of consumer electronics items. And, in recent years, hi-fi, stereo, videocassette players and recorders, wireless telephones, security systems, personal computers, compact disk

Fig. 3-3. This microwave dish, being moved into position by a helicopter, will be fastened in place by the workers shown here. COURTESY OF ANDREW CORPORATION.

(CD) players, camcorders, and other equipment have helped to keep the U.S. economy and the electronics industries of the U.S. and other countries healthy and active. This should continue for some time, at least, due to electronics newcomers such as multi-image TVs, stereo TV, high-definition TV, FMX (FM with greatly reduced noise and greatly expanded range), AM stereo, and the new DAT format.

Electronics engineers, technicians, repairers, and other specialists will be needed in great numbers during future years to design, manufacture, test, adjust, troubleshoot, demonstrate, and repair this equipment (Fig. 3-4).

Fig. 3-4. Technicians and engineers must exercise caution at all times when working on live circuits. (COURTESY OF NESDA/ISCET)

GOVERNMENT

The governments of most countries make extensive use of electronic equipment of many types and need workers to install, test, repair, adjust, maintain, and replace this equipment. Electronic equipment is used to create and store the huge numbers of records developed daily by the various departments and division of the government. Local, state, and federal governments use electronic communications systems, crime detection and deterrant equipment, space exploration instruments and controls, satellite communications, security and surveillance systems, and others.

The operation of electronic equipment repair depots is sometimes also the responsibility of federal or regional governments. At such a depot, radio transmitters and receivers, telephone switchboards, video systems and others are brought in and repaired, rebuilt, and tested as shown in Fig. 3-5, and sent back into the field for reuse.

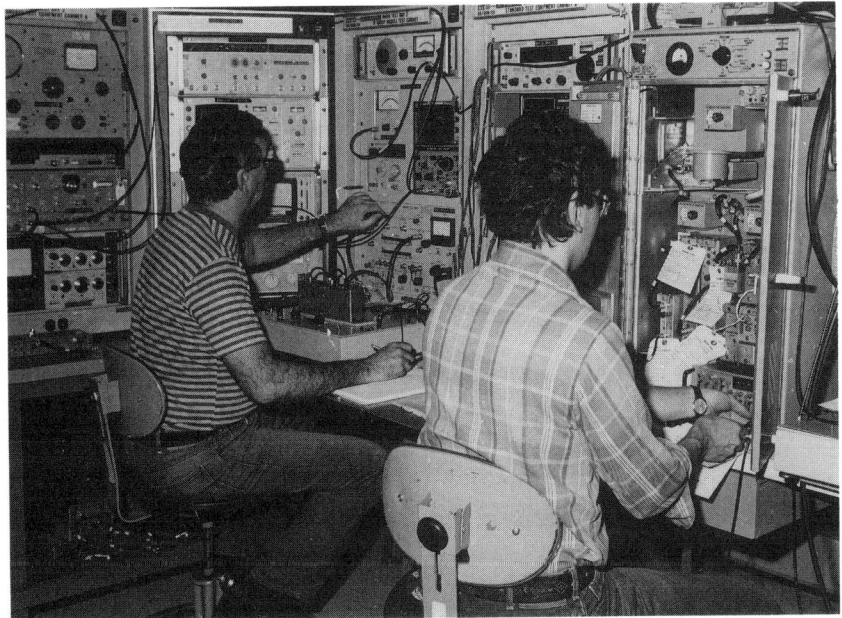

Fig. 3-5. These government employees in a repair depot are using special equipment to check out a receiving system. The technician on the left is checking a receiver modulator; the one on the right is testing a radio receiver. COURTESY OF TOBYHANNA U.S. ARMY DEPOT.

In the U.S., the Federal Communications Commission, the Federal Aviation Administration, the Department of Defense, public utilities commissions, and state police employ electronic engineers and technicians.

MILITARY

In many cases, the various branches of the military employ civilian electronics workers like those in Fig. 3-6, to adjust, install, repair, and update electronic systems and equipment. Also, military personnel with knowledge and skills in electronics are required in large numbers. Many of these people have little or no knowledge of electronics when they enter the military, but are trained in electronics in specialized courses to become electronics specialists. They install, maintain, and repair fire control radar, sonar, tracking, surveillance, and electronic instrument systems of many types.

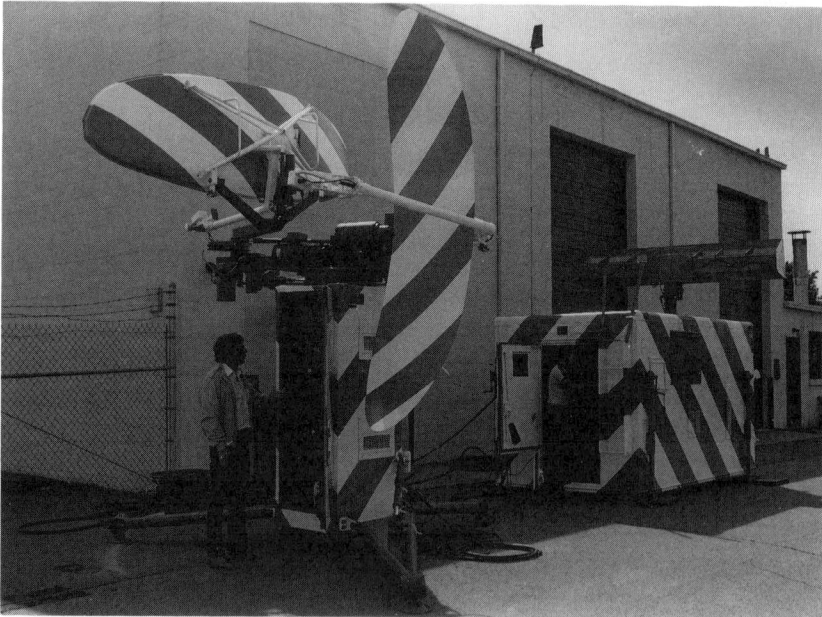

Fig. 3-6. Technicians are preparing to test a military radar system. COURTESY OF TOBYHANNA U.S. ARMY DEPOT.

❖ 4

Industrial
Electronics Positions

ANY OF THE ELECTRONICS POSITIONS AVAILABLE IN INDUSTRY
M are with manufacturers of electronic products and manufacturers of non-electronic products (such as motorboats, trac-
tors, or prefabricated houses). Most industries have greatly increased their uses of automation and robotics for producing and
processing their products. This has increased the need for personnel with electronics abilities.

Manufacturers of electronic products (industrial controls, TVs,
video-cassette recorders (VCRs), or radio transmitters) generally
employ a greater percentage of electronics people. These people
include electronics circuit and equipment design engineers (FIG.
4-1), electronic assemblers and production supervisors, quality
control checkers, testers, troubleshooters, repairers, and factory
engineers who can guide clients in the choice and use of electronic products and equipment.

Most jobs in industrial electronics are well-paying jobs with
good benefits and opportunities to advance. However, one impor-

tant requirement is that electronic equipment must be reliable and kept in operation to the fullest extent possible, with "down time" held to a minimum. Therefore, the manufacturer must see to it that the customer is sold the equipment that will do the job efficiently, that there is a minimal possibility that it will fail or break down, and that it can be fixed quickly by capable people if it does malfunction.

Another requirement of workers in industrial electronics is that the people who operate electronically-controlled equipment must be well trained. Some of these workers come from technical institutes.

ROBOTICS

One of the most interesting and fastest growing areas of electronics in industry is robotics, based on the use of machines called robots that perform much like humans. Today's industrial robots perform assembly routines that need superhuman strength, or

Fig. 4-1. A design engineer must be an expert with test instruments such as this oscilloscope, which displays several signals at the same time. COURTESY OF TEKTRONIX, INC.

36

are hazardous or tedious or very exacting. Factory robots may range from simple types that automate certain operations through progressively sophisticated ones that have been given the ability to touch, see, and hear. In many of these applications, robots perform their jobs better, quicker, and with greater precision than humans doing the same work. Robots are being used in relatively large numbers in the automotive industry and also in the assembly and manufacture of appliances, aerospace equipment, and various consumer goods. Their use is also being introduced in food processing, and in the textile, pharmaceutical, furniture, health care, and construction industries.

Robots are controlled to a great extent by electronics. Technicians who keep robots working, and make repairs when they break down, need a good knowledge of electronics. Usually they must also have knowledge in other technical areas such as pneumatics, hydraulics, mechanics, and optics. A good electronics technician can take local college courses in these topics, or study correspondence school lessons to learn most of what he or she needs to know. Also, robotics manufacturers sometimes provide either factory or on-site training in their overall robotics systems.

A breakdown in a robotics or automated production systems must be fixed quickly to keep down time and resultant financial loss to a minimum.

ENGINEERS

Electronics engineers are, in general, among the best paid. A 1987 salary survey made by the Institute of Electronics Engineers (IEEE) indicated that engineer members of the IEEE who were working as engineers in their area of expertise were earning an average wage of $53,889. This was an increase of 6.6% over the average salary in a similar survey of 1985. Of course, salaries for engineers can vary widely, depending on their age, the type of work they are doing, their education and degrees held, how long they have been on the job, and the specialized area of electronics they are working in. Electrical and electronics engineers make up the largest branch of engineering, holding over 390,000 jobs in 1984. Many of the jobs (in fact, most of them) were with firms making electrical and electronic equipment, scientific equipment, aircraft and aerospace equipment and parts, and business machines.

According to the 1986-87 *Occupational Outlook Handbook* published by the U.S. Department of Labor, electrical and electronics engineers should enjoy good employment opportunities through the mid-1990s. Employment is expected to increase faster than for all other non-electronic occupations. Increased demand by business, industry, and government for electronic equipment such as TVs, computers, communications equipment, and military electronic gear will account for much of the increased employment among manufacturers. However, the consumer market is using more and more electronics items such as security alarms, stereos, and wireless telephones, and there is an increased need for engineers to design and develop robots and other automation equipment. Some positions for electronics manufacturing engineers are in the defense industries, so budget cutbacks sometimes cause layoffs among engineers as well as other personnel. A new job for a laid-off engineer in the present times is typically not hard to find, but it might require relocating to a new geographical area.

SALES ENGINEERS

Many manufacturers of electronic equipment and systems employ engineers in sales positions, especially if the equipment or systems are high cost items such as radio transmitters or industrial robots. A salesperson who is also an engineer has the advantage of being able to discuss the special technical features of a product with the engineers and technicians employed by the customer. Often, these customer technicians and engineers influence heavily which products the employer purchases, especially where the products are highly technical in nature and where maintenance and troubleshooting and low down time are important considerations.

In many cases, an engineer whose main responsibility is design of the electronic product is taken along on sales calls by a salesperson for technical backup. A design engineer who turns out to be especially valuable in sales assistance frequently can move into the sales department, with higher pay the usual result. A good sales engineer often makes more money and can advance higher and more rapidly than an engineer whose main responsibility is design. However, design positions also pay well, and many engineers are not especially attracted to sales or management, preferring to remain in "pure" engineering positions.

PRODUCTION TEST TECHNICIAN

An example of one of the many types of jobs for technicians in industry is that of production test technician. In this position, the technician makes a specified group of tests on each manufactured or processed item. If all tests are within acceptable ranges, the item is then passed on to be packaged and approved for sale, or it proceeds to the next series of manufacturing or processing steps, followed by other specified tests. If the item does not pass the limits of one or more tests, it is shunted off to the side for another technician to find and repair the trouble, after which it goes back to the production line test technician. The work of the production test technician is routine; it's pretty much the same all day long. But it's an important job. This job often determines the reputation of the manufacturer as far as quality and reliability of product are concerned, so, while the work may be routine, the alertness and concern for product quality are great.

Production tests are also sometimes done by one or more of the engineers who designed the electronics system. This is to ensure that the overall system as a whole is working within the limits required for the application of the equipment.

QUALITY CONTROL

In manufacturing, the quality of products being produced or assembled or processed on the production line, or at the end of the final test, is typically spot-checked by a quality-control technician or engineer. In this position, the electronics specialist inspects items at certain intervals—such as every fifth, tenth, or twenty-fifth item—and makes a test of either certain characteristics or of its overall specifications. If the quality control person detects problems, he reports it to the foreman or supervisor who isolates the cause of the problem. If a lapse of quality occurs at a rate that is considered excessive, it is not unusual for the foreman to stop the whole production line and check every item. In addition, of course, he must determine the cause of the problem and correct it.

Quality control work is generally not as routine as production test work because in quality control, it is essential to determine the cause of the reduced quality. It could mean that the wrong person is in some production-line job — or that some person is in a wrong job which is almost, but not quite, the same thing. Of course, there could be any or a number of other reasons; for instance, the quality of the components or materials being used has deteriorated, the wrong solder or wrong solder temperature is being used, or the line is being moved faster than the workers are capable of doing a good job.

DISTRIBUTORS AND REPRESENTATIVES

Some of the larger electronics distributors and manufacturer's representatives employ technicians and sometimes engineers to advise clients how to use and operate electronics goods, or to help the client select an item of electronic equipment for a particular application. They also often make repairs or corrections on equipment not operating properly, or help to clear difficult problems encountered by the customers' technical personnel. Their job is sometimes also to check out an electronic system or equipment ordered by a customer prior to installation at the client's premises. It is not unusual for electronic equipment to become slightly out of adjustment between the time it leaves the manufacturer and when it arrives at the distributor. Advanced checkout can head off later complaints.

ELECTRONICS ASSEMBLERS

One of the largest categories of electronics workers is that of electronics assembler. If you start out as an assembler, your pay might be somewhere around the legal minimum wage or a little above. But some assemblers make more. Especially if their assembly work requires more skill, care, or ability than average. Assemblers perform work at a level less than that required of the precision level. Usually included in the assembler classification are electronics wirers. These are the people who interconnect parts on circuit boards by soldering wires, assemble cables and connectors, and operate wave-soldering machines which are devices that make dozens of connections all at once in a dipping or wave operation.

Most entry-level assembly jobs require comparatively little formal education — usually only a high-school diploma, although some employers waive even that requirement. The ability to perform efficiently on the job is the main requirement. The training needed for the work usually takes only days or a week or two at the most.

There are some exceptions. For example, precision assemblers of electronic circuits and systems (FIG. 4-2) sometimes require technical or trade school training or the equivalent. These precision jobs provide a somewhat higher pay scale.

Assembler jobs are subject to employment fluctuations. When the economy takes a dip, factory orders slump, and layoffs occur. In general, the employment outlook for electronics assemblers for the foreseeable future is good. According to the U.S. Department of Labor, there were 353,000 precision assembly workers of all types in the U.S. in 1984, and 41 percent of these worked in industries where electronic or electrical machinery and equipment are manufactured. It is estimated that the number of assemblers needed will increase slowly each year up through the mid-1990s. While work as an electronics assembler might not be the best long-term electronics career in some cases, it's a good way to "get your feet wet" in electronics. It's possible to advance from assembler to other more responsible jobs at better pay, such as lead assembler, quality checker, production tester, foreman, supervisor, and so on.

The average earnings for all assemblers working full-time in 1984 was $291 a week. Precision assemblers earned 5 percent to 10 percent more. Precision assemblers working under union contract earned $8.50 to $15 per hour according to some estimates. Some of the workers earning these higher wages are in the defense and aircraft industries. In some cases, wages are tied to incentive or piecework rates. The more items you assemble or assembly operations you complete, the greater your pay at the end of the week.

Another type of assembly job is that of development electronics assembler. Usually an electronics technician is assigned to the job and a high degree of skill and training is required. The equipment assembled is sometimes an electronics prototype circuit and the assembly is done from a schematic. The need to make and record tests after assembly is completed is often required. The job of a

Fig. 4-2. Today's electronic products are manufactured using a combination of manual and automated assembly testing techniques. COURTESY OF JOHN FLUKE MFG. CO.

developmental electronics assembler can be very interesting and the wages earned are also usually higher.

AEROSPACE AND AVIATION ELECTRONICS

Electronics technicians and engineers employed in the aerospace and aviation fields design, develop, test, produce, install, and repair communications and other electronic equipment used on the ground, in the air, or in space. Their work often involves electronic equipment in civilian and military aircraft, missiles, and space vehicles. They are sometimes involved in developing new technologies in aviation, defense, and space electronics.

Most of the electronics personnel working in the aerospace field are employed by National Aeronautics and Space Administration (NASA). Others are employed by the FAA (Federal Aviation Administration), the manufacturers of aerospace and communications equipment, airlines, and companies specializing in contract work on installing, servicing, and repairing this equipment.

More information on the aerospace program and employment in aerospace electronics can be obtained from NASA (See Appendix F for address.)

Information on open positions with the Federal Aviation Administration (FAA) are often published by associations such as the Electronics Technicians Association. The following example of types of positions available with the FAA is quoted from the POSITIONS AVAILABLE section of an issue of the ETA Technician News sent regularly to ETA members:

FEDERAL AVIATION ADMINISTRATION
ELECTRONICS TECHNICIANS—The FAA employs thousands of Electronic Technicians for the following career fields—Facility Installation (FE), Navigational Aids, Communications, Environmental, Radar, and Automation. Electronic Technicians install and maintain electronic equipment systems required for aerial navigation communications between aircraft and ground services, and control of aircraft movements to assure safety in air traffic. This involves work with radar, radio, transmitters/receivers, computers, and other devices at airport, en route traffic control centers, and facilities along

Table 4-1. FAA Regional Offices Which List Job Openings. (COURTESY OF
ELECTRONIC TECHNICIANS ASSOCIATION, INTERNATIONAL.)

HEADQUARTERS: Federal Aviation Administration (AHR-150), 800 Independence Avenue, Southwest, Washington, DC 20591, (202) 267-8007
ALASKA REGION: Alaska and Aleutian Islands
Federal Aviation Administration (AAL-14), 701 C Street, Box 14, Anchorage, AK 99513, (907) 271-5747
EASTERN REGION: Delaware, Greenland, Maryland, New Jersey, New York, Pennsylvania, Virginia, and West Virginia
Federal Aviation Administration (AEA-14), JFK International Airport, Jamaica, NY 11430, (718) 917-1060
CENTRAL REGION: Iowa, Kansas, Missouri, and Nebraska
Federal Aviation Administration (ACE-14), 601 East 12th Street, Kansas City, MO 64106, (816) 374-3304
GREAT LAKES REGION: North Dakota, South Dakota, Illinois, Indiana, Minnesota, Michigan, Ohio, and Wisconsin
Federal Aviation Administration (AGL-14), 2300 East Devon, Des Plaines, IL 60018, (312) 694-7731
NEW ENGLAND REGION: Connecticut, Maine, Massachusetts, New Hampshire, Rhode Island, and Vermont
Federal Aviation Administration (ANE-14), 12 New England Executive Park, Box 510, Burlington, MA 01803, (617) 273-7345
NORTHWEST MOUNTAIN REGION: Colorado, Montana, Utah, Idaho, Oregon, Washington, and Wyoming
Federal Aviation Administration (ANM-14), 17900 Pacific Highway South (C-68966), Seattle, WA 98168, (206) 431-2014
SOUTHERN REGION: Alabama, Florida, Georgia, Kentucky, Mississippi, North Carolina, South Carolina, and Tennessee
Federal Aviation Administration (ASO-14), P. O. Box 20636, Atlanta, GA 30320, (404) 763-7706
SOUTHWEST REGION: Arkansas, Louisiana, New Mexico, Oklahoma, Texas, and Mexico
Federal Aviation Administration (ASW-14), P. O. Box 1689, Fort Worth, TX 76101, (817) 624-5014
WESTERN-PACIFIC REGION: Arizona, California, Nevada, Hawaii, Pacific Ocean area west of continental United States (except the Anchorage, Seattle, and Oakland Flight Advisory areas), and east of East Pakistan and India-including all free nations south and east of China
Federal Aviation Administration (AWP-14), P. O. Box 29007, Worldway Postal Center, Los Angeles, CA 90009, (213) 297-1305
OTHER MAJOR FIELD OFFICES: Federal Aviation Administration, Technical Center (ACT-14), Atlantic City, NJ 08405, (609) 484-6621

Federal Aviation Administration, Mike Monroney Aeronautical Center, (AAC-14), P. O. Box 25082, Oklahoma City, OK 73125, (405) 686-4506 FAA regional offices to which inquiries should be directed on job openings for electronic technicians and engineers.

the network of Federal Air Traffic Control routes. Job responsibilities include preventive maintenance, replacement of deteriorating parts, and corrective maintenance on malfunctioning electronic equipment. Electronic Technicians may also specialize in design, development, and evaluation of new types of electronic equipment and the National Air Space Systems (NAS). An Equal Opportunity Employer M/F

Inquiries on these positions and others available with the FAA should be directed to the Employment Office serving the geographical areas listed in TABLE 4-1. (This listing was also part of the information provided in the ETA notice).

❖ 5

Maintenance and Troubleshooting in Industry

I NDUSTRY PROVIDES POSITIONS IN SEVERAL AREAS, INCLUDING PRO-
duction electronics, instrumentation, factory and field service,
repair, and biomedical electronics.

PRODUCTION ELECTRONICS

While most of the troubleshooting and maintenance work done
in industrial production is done by technicians, some is done by
engineers, and some of the routine maintenance is done by repairers and apprentices. If electronic equipment is what is being
produced, electronic technicians are often employed in the manufacturing or production processes. These technicians are usually responsible to engineers or scientists, but much of their work
is done without close supervision. In many cases they supervise
workers on the production line to ensure that the products are

being manufactured according to specifications (FIG. 5-1). In some cases, the technicians specify the parameters for the products being made and write up or conduct tests on the final products.

Sometimes research and development work is assigned to production technicians. In a typical case, the technician sets up the equipment called for by the engineer or scientist, prepares the experiments, and uses mathematics in addition to instrument measurements to determine the results. Research and development technicians sometimes assist engineers in the design of circuits and equipment, build prototypes of the new designs, and evaluate the performance of the circuit or equipment.

INSTRUMENTATION

Electronic instrumentation technicians are employed by industry, the government and the military to install, test, repair, maintain, and adjust indicating, recording, controlling, and tele-

Fig. 5-1. Some test functions in a manufacturing process are no longer manually checked individually. Here, printed circuit boards are tested before and after assembly on automated test systems. COURTESY OF JOHN FLUKE MFG. CO.

metering instruments of various types used to measure and control variables such as flow, pressure, motion, temperature, force, and chemical composition (FIG. 5-2). An instrumentation technician normally must have an ability with math, and a good knowledge of basic science and the fundamentals of physics.

Many electronic instrumentation technicians and engineers are employed in the petrochemical industries; they work for petroleum production and pipeline industries. Others work in the medical field (FIG. 5-3), in laboratories, for pharmaceutical manufacturers, in processing industries, in space and satellite electronics, and in many other activities where controlling, monitoring, and measuring are done. There are approximately 60,000 electronics repairers and others who are employed fulltime in repairing the instruments used in the field mentioned. When you include other technicians and engineers who also install, adjust, design, or otherwise are active in industrial instrumentation, their number is considerably greater. Technicians working in the industrial field are paid from $15,000 to $25,000 and upward.

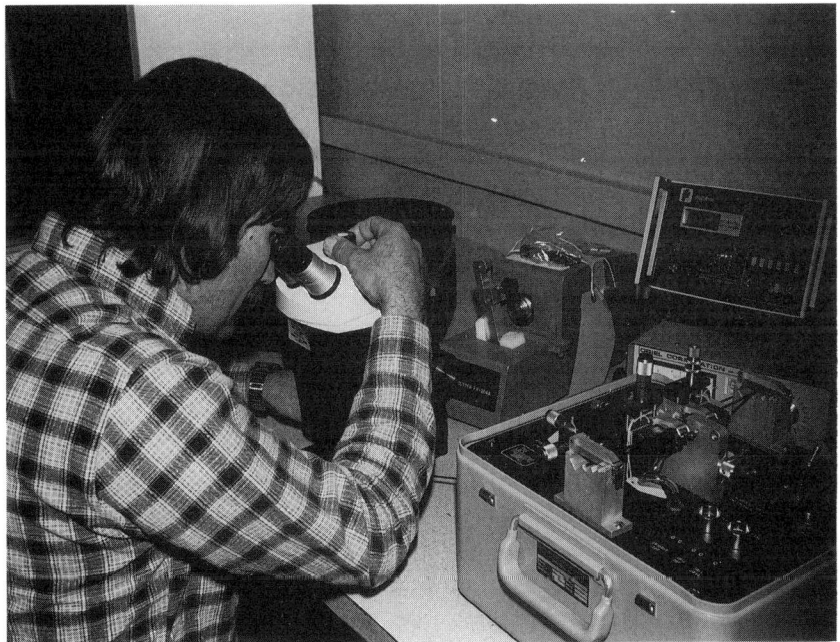

Fig. 5-2. Many technicians employed by the government work on state-of-the-art electronic equipment. The technician here is using a microscope to inspect a splice that has been made in a fiber-optic cable. COURTESY OF TOBYHANNA U.S. ARMY DEPOT.

FACTORY SERVICE AND FIELD SERVICE

Manufacturers of electronic equipment often employ engineers and technicians who spend much of their time away from the plant (FIG. 5-4). They install, adjust and repair equipment being used by the manufacturer's customers. Sometimes they are tech school or college instructors who also conduct training sessions teaching other technicians and engineers how to select, apply, operate, install, troubleshoot, and repair the equipment (FIG. 5-5). Factory service and field service engineers and technicians represent the manufacturer and are generally thoroughly trained on specific products. Sometimes they make product tests in the factory. In some cases, their work is to keep electronics equipment or systems operating within specs during a warranty period. In other cases, after the warranty period has run out, they can be available at an hourly or daily rate to locate the problem and put equipment or a system back in operation. Some factory engineers and technicians remain at the manufacturers plant and

Fig. 5-3. Many electronic technicians and engineers work in the medical field. COURTESY OF TEKTRONIX, INC.

are available by telephone for consultation on difficult problems (FIG. 5-6). Oftentimes, if the problem cannot be solved by telephone, the factory electronics representative can be contracted to travel to the customer's location to assist in clearing the trouble.

Some owners of electronic equipment such as photocopy machines and computers have a service contract with the dealer,

Fig. 5-4. An electronic technician or engineer who works in customers' offices, homes, and businesses should be mannerly and be diressed in a clean, pressed uniform or shirt and tie.

Fig. 5-5. College instructor explaining master-antenna distribution system to technicians. COURTESY OF ELECTRONIC TECHNICIANS ASSOCIATION, INTERNATIONAL.

distributor, or manufacturer. The service contract provides for a regularly scheduled visit by a field service technician or field engineer (also sometimes called customer-service engineer) to clean, readjust or realign, and completely recheck the equipment, putting it back into practically factory-new condition. Engineers and technicians who call on customers at their residences or business locations must be capable of dealing with clients in a professional manner and be dressed in business suit, uniform, or other presentable attire. They spend much of their time listening to customers' complaints, explaining the technical workings of equipment, offering advice and operation of the equipment, discussing the operation of particular circuits, and so on. In effect, a technician or engineer in such a position is a public relations representative for his employer.

REPAIR

Many of the jobs available in industry have the words "electronic repairer" or "repairer" as part of the position title. Some

Fig. 5-6. In some cases, manufacturers factory-engineers can assist by telephone in solving a problem on equipment. COURTESY OF TEKTRONIX, INC.

repairer jobs require somewhat less than a 2-year trade or technical school education in electronics, but with specialized training in a particular area of electronics. Such positions might include office machine and cash register service, TV and radio repairer, audio/video repairer, public address repairer, and electronics mechanic. In some repairer jobs, a high degree of skill on a particular product or part of a system is required. Training on the product or system is typically provided by the employer, so even though the educational requirements are sometimes less than those often specified for technician jobs, an electronics repairer frequently earns excellent wages, especially those who work for a particular manufacturer.

Electromedical or biomedical equipment repairers work in the field for a manufacturer of certain equipment or class of equipment such as x-ray machines. In this work, the repairer tests, adjusts and repairs the equipment quickly and efficiently, since he has concentrated training on that item. Pay scales for repairers may range from $12,000 starting to $25,000 or more. Advancement to higher level jobs is possible.

BIOMEDICAL ELECTRONICS

Other personnel who work on biomedical equipment often are required to have at least two years of technical school education and training in electronics. In particular, the biomedical electronics technicians who work for hospitals usually require substantial hands-on training on that type of equipment which normally is part of a 2-year associate-degree electronics program which probably would include serving an internship in a hospital. Some hospitals are starting to prefer graduates of 4-year degree programs in electronics or specializing in biomedical electronics. Other hospitals require certification by a biomedical electronics association within a certain period of time after being employed.

Many biomed electronics technicians and engineers are hired at somewhere around $15,000 and can earn up to $40,000 and more, with excellent chances for advancement to positions of greater responsibility. These jobs require a professional appearance and manner since the work requires close contact with patients, nurses, doctors, and other professionals.

❖ 6

Employment in Broadcasting and Cable TV

T ECHNICIAN OR ENGINEER JOBS IN THE FIELD OF RADIO OR TV
broadcasting and cable TV are among the more difficult electronics positions to obtain because so many people apply for those
positions. However, openings do occur. Knowing about them
when they occur, and being in the right place at the right time,
makes it reasonably possible to be hired into one of those jobs. In
broadcasting and cable, as in many other industrial applications
of electronics, an engineer does not necessarily have a college
engineering degree. Often the person who is responsible for the
technical equipment of a radio or TV station and its proper operation, and for repairing it, has the title "chief engineer." Among his
other qualifications, the chief engineer is ideally a technical person who knows how all the equipment works, how to test and

evaluate it, how to make sure it is operating according to FCC regulations, and how to repair it.

RADIO BROADCASTING

In radio broadcasting, the chief engineer must frequently be on call or available 24 hours a day, or, he must have one of his other technicians or engineers available at all times in case a technical problem occurs. At a great many small radio stations, the chief engineer is the only technical person on the staff, so he is the one who takes care of anything that comes up. At some of the smaller stations, he acts in a dual role: chief engineer and also part-time announcer, news person, or disk jockey. To get a start into broadcasting, especially with radio, it could be to your advantage if you have a good speaking voice.

It is also to your advantage to have an FCC license, even though the FCC has recently deregulated radio and TV broadcasting. The FCC no longer requires that broadcast engineers and technicians have a license in order to operate and adjust transmitting equipment. The FCC now leaves that up to the station licensee—that is, the owner of the station. Generally, however, the licensee, or the general manager acting for the licensee, tries to hire engineering people who have such a license or equivalent certification as insurance that its technical staff is competent.

Some small radio stations require the services of a technical person only on a part-time basis. As a result of this, some engineers work for two, three, or more radio stations, having the title of chief engineer or operator at each one. These technicians or engineers stop in at each station every day or several times per week to check on the operation. They also spend about one night per week at each place on preventive maintenance or making repairs on equipment that has begun to deteriorate in operation. Of course, in case of a breakdown at any station, the engineer is available to be called in on a moment's notice to get the station back on the air as soon as possible.

TELEVISION BROADCASTING

Television broadcasting, like radio broadcasting, no longer requires an FCC license; it's up to the licensee (station owner) to hire qualified technicians and engineers to install, operate, and maintain equipment. (A production department engineer is shown on the job in Fig. 6-1.) The station manager, and chief engineer are accountable to the FCC to see that the station is run properly. Thus, as in radio broadcasting, TV broadcasters generally hire technicians and engineers who are licensed or certified or both. Commercial TV stations generally charge more for their advertising than radio stations and therefore lose more money when a breakdown causes them to "go dark" (go off the air). Thus, people who repair, troubleshoot, and maintain TV broadcast equipment — especially those who are on call in case of trouble — must be well-qualified and able to do their work quickly and properly.

Most technicians and engineers at TV stations are graduates of resident technical schools, having completed at least a one- or two-year course. Some are graduates of schools offering specific training in broadcast operation and engineering. Some are also graduates of good correspondence courses aimed at broadcasting and technician licensing.

Fig. 6-1. Broadcast production engineer looking over stored images available for use in a TV program. COURTESY OF AMPEX CORPORATION.

INSTALLATION, MAINTENANCE, AND TROUBLESHOOTING

To be responsible for the installation, maintenance, or trouble-shooting of radio and television station and cable TV equipment, you need not only be familiar with electronic circuits and how they work, and what the equipment is supposed to do; you must also be highly skilled in wiring and soldering and the use of hand tools and—most of all—in the use electronic test instruments, like the ones in FIG. 6-2.

The test instruments used nearly always include oscilloscope (sometimes a sophisticated one), the volt-ohm millimeter (analog and digital), frequency meter, radio-frequency generator, distortion meter, and modulation meter. For TV broadcasting, and for technical cable TV work, you must also be skilled with the use of the spectrum analyzer, wavemeter, waveform monitors, phase monitor, sweep generator, side-band analyzer, video generator, color test and monitoring equipment, and other test instruments.

You must be also on intimate terms with the detailed characteristics of the composite video signal and the video-modulated carrier signal. There are standards set by the FCC, the EIA, and

Fig. 6-2. TV broadcast engineers use complex test and measuring instruments every day on the job. COURTESY OF TEKTRONIX, INC.

the National Cable Television Association (NCTA) that a good video signal must adhere to. These standards set limits for such characteristics as frequency, rise time, fall time, width, frequency response, overshoot, and droop. Knowing how to use a precision oscilloscope and other instruments for measuring these characteristics is essential.

MAINTENANCE SUPERVISOR, CHIEF ENGINEER

In most TV broadcast stations, there is usually a maintenance supervisor and a chief engineer. Sometimes the maintenance supervisor is also the assistant chief engineer. Both positions require people who are highly qualified and experienced in all technical matters. Usually, they are also knowledgeable in other areas of building maintenance. For example, if the air conditioning, heating, ventilation, intercoms, or telephone systems break down, it's often the maintenance supervisor or the chief engineer who's called first. He then evaluates the trouble and either has someone on his staff get at it, or calls an outside contractor and oversees repairs.

In larger TV stations, the maintenance supervisor is responsible for seeing that maintenance technicians are doing their work properly or for instructing them on how to do it if they are new on the job. He is usually next in line to become chief engineer.

The chief engineer of a TV station must also be able to communicate with other department heads as a member of the management of the station. He keeps the station manager informed on possible problems, on whether troubles have been corrected, when new equipment is needed, what the advantages and characteristics are of new equipment designs, the implications of any changes in FCC Rules and Regulations, and so on.

In dealing with the news director, the program director, and other program people, the chief engineer makes sure they know how to make the best use of the cameras and special effects and other equipment available to them and is able to advise them on new developments, such as using satellite transmission and reception (FIG. 6-3).

The TV chief engineer must sometimes work at night to ensure that major changes or repairs in equipment are made properly, or at a mountain-top TV transmitter when a new antenna or microwave dishes are being installed atop a 1500-foot tower or atop a

Fig. 6-3. Equipment racks for transmit/receive earth station handling video programming. COURTESY OF ANDREW CORPORATION.

building (FIG. 6-4). He must also be on hand to attend staff and management meetings. Being the chief engineer of a TV station is a tough but interesting job. It pays well and is within the reach of the qualified and well-rounded electronics engineer or technician.

TRANSMITTER ENGINEER

Some TV broadcast stations have their TV transmitters manned on a 24-hour basis; at least one qualified engineer is on duty at all times. This engineer observes the operation of the transmitter on a more-or-less continual basis. The levels of the video signal, the synchronizing signal, the black level, the white peaks, and the modulation percentage are checked using waveform monitors. A waveform monitor is a special purpose oscilloscope. He also checks the audio levels and the quality of the audio signal. He makes sure the tower lights are on when darkness is approaching, that the antenna heaters are working when icing conditions occur, that the cooling systems are keeping the high-power transmitter output tubes within safe temperature limits, and that aural (sound) and visual (picture) transmitters are operating at proper power.

Sometimes he also must check on microwave links when these are part of the transmitter equipment. In spare moments, he carries out preventive or corrective maintenance on equipment not then in use.

At some TV transmitters, the engineers work an 8-hour shift. The transmitter is usually located on the top of a hill or mountain, or in a tall building. If it's on a hilltop, the engineer coming on duty typically drives 5 to 25 miles or more to the transmitter building in a station-owned 4-wheel drive vehicle, sometimes with a snowplow attached for clearing the seldom-travelled road. When he arrives to take over the next shift, the person going off duty takes the same vehicle and drives back to the city where his own car has been parked at the TV studio or other location, leaving the 4-wheel drive vehicle there for the next shift.

Fig. 6-4. A TV chief engineer doesn't have to climb tall towers but must be able to arrange for and supervise installation and operation of microwave and satellite dishes and other such equipment. COURTESY OF ANDREW CORPORATION.

Sometimes, a transmitter engineer works more than an 8-hour shift. Sometimes he works a 24-hour day, or two or three days. In some cases, transmitter engineers work for a week or a month and then have an appropriate amount of time off before their next tour of duty. Where such long shifts are required, the transmitter building is usually comfortable and well-equipped with food and emergency supplies. In case of any unusual emergency, a helicopter can bring in help. Some TV transmitter positions require a special kind of person — one who is willing to spend a lot of time alone and who has few family obligations.

Positions in radio and television broadcast stations are often advertised in magazines such as *Broadcasting*, *Television Engineering*, *Radio World*, *TV Technology*, *Television Broadcasting*, and the SBE *Signal*. Certification as a broadcast technologist or as a broadcast engineer can be provided by the Society of Broadcast Engineers (SBE). Information can also be obtained from one of the local SBE chapters. (See Appendix J for addresses). You can also learn of openings for broadcast engineers and technicians, as well as get to know the people who hire technical broadcast people, by attending local SBE meetings. Ask to be put on their chapter mailing list, or join as a member and automatically receive meeting notices.

CABLE TV TECHNICIANS

Cable television companies also hire electronics engineers and technicians to install, adjust, repair, operate, and maintain their systems. These systems include multiple antenna installations, amplifiers, and cable distribution networks that distribute the received television channels to its hundreds or thousands of customers. Most cable TV companies also originate some programs and advertising of their own, much as a broadcast station does, and some have one or more channels of their own. Many also have satellite receiving terminals for receiving certain programs which subscribers sometimes get for additional cost.

Cable TV installers often work out doors installing subscriber drops from the main cable to individual homes or apartments. Some cable technicians work at satellite receiver locations or at antenna locations either fulltime or on a scheduled or periodic basis. A good knowledge of electronics and the ability to use test equipment is a necessity for this type of work. Information on employment in the cable TV industry is available from the National Cable Television Association and from the Society of Cable Television Engineers (SCTE). (Addresses can be found in Appendix G.)

WORKING CONDITIONS

Broadcast technicians who work at transmitter sites and who do work as part of news teams often must work out of doors in bad weather or in difficult situations. Cable TV installers also often work outside, connecting drop cables between customers' locations and the main cable line or trunk. Some need to learn to climb poles using spikes or to work on the pole from a ladder or from a cherry picker bucket. Most other cable and broadcast technicians work in pleasant and comfortable surroundings in a studio or cable head end and remote-broadcast van. Many times a technician must work at night, either on a regular schedule or on a rotating shift, since broadcasting and cablecasting go on 24 hours a day in most locations. Sometimes the work is under pressure. For example, it's sometimes necessary to get equipment set up in time for a remote program that's just been scheduled, or for an accident, flood, or other newsworthy event.

There are presently approximately 30,000 technicians and engineers employed in broadcasting and cable television in the larger towns and cities, most of them in television. Employment in the field is increasing faster than the average growth rate in other electronics fields. Most of the better jobs are in the largest cities.

EDUCATION AND TRAINING

The training required for technician and engineering work in broadcasting and cable TV is usually two years in a technical school or community college, or equivalent. Courses specializing in broadcasting and/or cable TV provide graduates with more of an edge in getting hired or promoted. If you are the graduate of a four-year college course in electronics or electrical engineering, this provides you with a better chance of getting hired or promoted. Some technicians have been hired into an entry level job on the basis of having completed a good correspondence-school course in electronics or broadcasting.

PAY SCALES

Latest figures from the U.S. Department of Labor on salaries in broadcasting and cable TV indicate that the average pay for technicians was about $330 weekly in 1984. Wages are somewhat higher than that now. The average is a little lower in radio and a little higher in television. However, both radio and TV technicians earn considerably more than the average if they work for larger stations in larger cities. Weekly salaries of $500 to $1000 per week are not too unusual in those larger stations and in technician and engineering positions with the networks and major cable channels. Chief engineers and supervisors sometimes earn even more. Positions in the larger stations are usually filled by people who have gained knowledge and experience while working in smaller stations.

It is often necessary to work overtime in broadcasting and on weekends and holidays. In those cases, pay at time and a half the hourly rate is usual, and sometimes double time. At some locations, membership in a union after 90 days of employment is either available or required for non-management employees.

7

Consumer Servicing

T ELEVISION SETS AND SMALL MODERN-DAY RADIO RECEIVERS MAKE
up a great portion of the electronic equipment purchased by
consumers. Technicians are still needed to service these, and
they make a good living at it in many cases. However, TVs and
small radios are now much more reliable than they once were so
they don't break down as often. It's not unusual for a small radio
or TV to be thrown away rather than to have it fixed when it stops
working. Often, the price a technician would have to charge to fix
a small radio would exceed the cost of a new one.

The fact that many households now own two, three, or four TVs
keeps the TV repair business still worth considering, but many
other electronic items that also warrant serious consideration of
consumer electronic servicing as a career. These include video
cassette recorders (VCRs), CB radio, home security and fire-alarm
systems, stereos, automobile radios, home computers, video
games, compact disk (CD) systems, video cameras, home satellite
earth stations, and home telecommunications (telephone) systems.

LEARNING THE TRADE

The Electronic Industries Association (EIA) provides schools with guidelines and course outlines for the training of consumer electronic service technicians. Most larger cities have schools that teach consumer electronics servicing. Some technicians get their start in consumer electronics servicing through study of correspondence courses, some of which include kits or training aids providing hands-on experience as the lessons progress. This is a good way to learn, especially if it is necessary to continue working as you study, or if you travel on your present job, or if a resident school is too far away to get to on a daily basis. Learning to repair radios, TVs, and other consumer electronic equipment can probably be learned more effectively in a resident trade or technical school. Membership in a technicians association is a good way to keep up to date on latest developments (FIG. 7-1).

Fig. 7-1. Members of technician association observing effects of adjusting earth-station antenna. COURTESY OF ELECTRONIC TECHNICIAN'S ASSOCIATION, INTERNATIONAL.

SCOPE OF EMPLOYMENT

From 1980 through 1983, the number of workers in consumer electronics, including manufacturing, fell from 84,200 to 66,800, according to EIA. But in 1984, the number employed rose to 70,500. The decrease was probably due to more TVs and other equipment being made in foreign countries. The increase starting in 1984 is due partly to an improved economy and to increased sales of new electronic products such as VCRs, compact disks, home computers, satellite reception systems, and so on.

Present annual employment in the total U.S. electronics industry is 1,800,000.

PAY SCALES

According to the International Society of Certified Electronics Technicians (ISCET), wages of their service technician members for the year 1984 ranged between $3.50 per hour for an entry level technician to $15.00 per hour for an experienced technician. By now, the average hourly wage is somewhat higher.

WORKING CONDITIONS

Technicians working in consumer electronic service work are sometimes bench technicians (FIG. 7-2), who are seldom seen by the public. Others call at individual homes and apartments (FIG. 7-3). For them, appearance, manner, and attitude — in addition to technical knowledge and ability — are all-important. An "outside" service technician should be neatly dressed, and should be especially careful of customers' furniture and other property and belongings.

GETTING YOUR FIRST JOB

A typical way to start in consumer electronics servicing is to go to work, perhaps part-time, for a small shop. You can then learn a variety of the types of work that a service technician is called on to perform. In a small shop you might have to start out as a part-timer. As you gain experience, you'll probably find that your hours are increased. However, if advancement seems limited, you can look around for work in a larger shop, or even consider opening your own electronics service business.

More information on the consumer electronics service business is available from ISCET and from ETA. (See Appendix G for addresses.)

ADVANCEMENT

As a service technician repairing consumer TVs and other equipment, advancement in pay and responsibility is possible by becoming a service manager in one of the larger shops, or by being assigned to the "tough dogs" (the sets whose troubles cannot easily or quickly be located and repaired). In smaller shops, tough dogs are handled by the owner in the example in FIG. 7-4.

Advancement as a service technician is also possible as a technician working for a local distributor of certain brands of TV sets, stereos, or VCRs. TV receiver distributors usually employ a num-

Fig. 7-2. The instrument this technician is using is for converging the three beams in a cathode-ray tube to provide the best color on a TV screen.
COURTESY SENCORE ELECTRONIC TEST INSTRUMENTS.

Fig. 7-3. Problems that cannot be fixed quickly in the home should be brought to the shop. If the set has a tilt-out drawer like this one, in-home repairs can be done more quickly. COURTESY OF NESDA/ISCET.

ber of technicians having different responsibilities. One technician handles telephone calls from shop service technicians who are having problems with or who need more information on a certain model of that distributor's TV or other equipment. Another technician takes and fills orders for parts. Another might have the responsibility of handling special problems with consumers. And another might conduct seminars, training sessions, keep track of modifications improving receiver performance, and so on. Service jobs with distributors sometimes combine two or more of these responsibilities.

Positions with larger organizations usually include improved benefits such as health insurance, vacations, and holidays with pay.

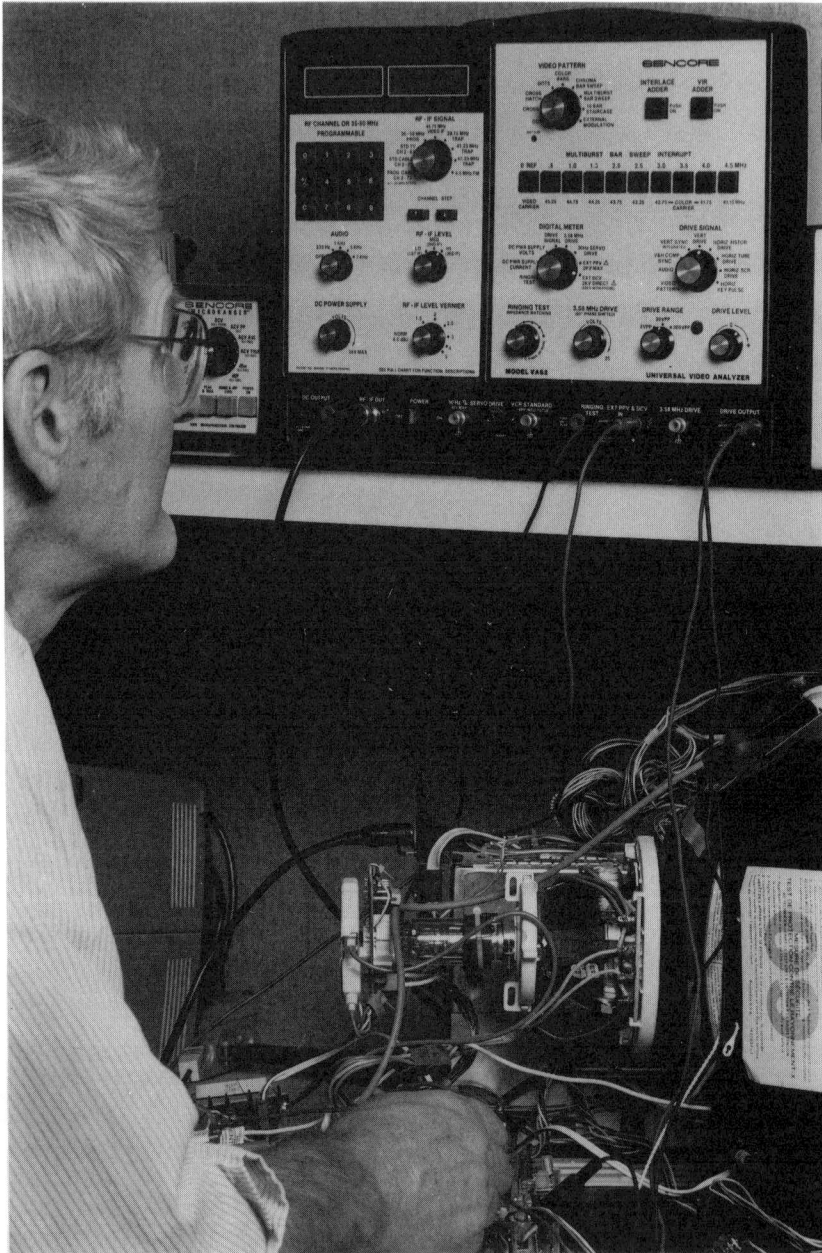

Fig. 7-4. This technician is using a complex but versatile instrument, a Universal Video Analyzer, to feed a known-good video signal to a TV set.

8

Telecommunications

T ELEPHONE COMPANIES STILL DO WHAT THEY HAVE DONE FOR MANY years: they provide a telephone service for individuals and businesses. But, now they also do many other things. At one time all of the Bell companies and the telephone manufacturer, Western Electric Company, were both owned by American Telephone and Telegraph Company (AT&T). Recently, however, the Federal Communications Commission (FCC) ordered a breakup of this organization. The Bell companies are now on their own, some having joined together to form new telephone companies serving two or more states (such as Bell Atlantic). Western Electric is still a manufacturer of telephone equipment. AT&T is still in business on its own, selling telephone equipment and providing long-distance and other services.

Aside from the Bell Companies, many so-called independent telephone companies existed. Many of these were members of the United States Independent Telephone Association (USITA). Since the AT&T breakup, virtually all telephone companies in the U.S. are now independent telephone companies, and a new organization has been formed called United States Telephone Association

(USTA). Information on which telephone company or companies serve your area can be obtained from USTA. (Their address is in Appendix G.) Of course, you can look in the local telephone directory or simply at your monthly telephone bill to get the address of your local telephone company. Also, the Yellow Pages of the telephone directory provide the addresses of any other nearby telephone companies.

USTA membership is available only to telephone companies, not to individuals. One of the functions of USTA is to provide or recommend training of their members for employees in both technical and non-technical areas. Some of these courses are also available to interested persons who are not necessarily employees of a telephone company. For further information on the courses, write to the Engineering and Technical Disciplines Department of USTA.

While long-distance service is still being provided by AT&T, some new companies have also come on the scene to offer long-distance service, and the subscriber has a choice of long-distance companies. Other companies include GTE Sprint, MCI Telecommunications Corporation, ITT Long Distance, and C-TEC Corporation. Addresses of local offices are listed in the Yellow Pages of your telephone book or are available from the reference section of a nearby public library.

CHANGES AND TRENDS IN TELECOMMUNICATIONS

In recent years, telephone companies have gotten into other business, and some new companies have gotten into the telephone business. These activities are classed under the general term *telecommunications*.

Some telecommunications activities include conventional telephone service, the transmission of data between two points or among a number of points, the use of computerized methods for more rapid or more efficient transmission, transmission of pictures by TV or facsimile, and transmission of voice and data simultaneously. Transmission of various types of information may be by any of several means. While conventional wire lines are still used, there has been increased use of radio and microwave transmission (FIG. 8-1), and recently transmission has also been by means of communications satellites for medium and long distances and by fiberoptics for shorter distances.

COMMUNICATIONS COMPANIES

There has been a considerable expansion in the number and types of companies that provide communications services for the public and for commercial and industrial organizations. Some of these organizations transmit medical data including x-rays and cardiograms between hospitals and doctors. Some transmit data between one industry or location and another, between banks, or between libraries. Others provide a means of electronic shopping; customers view an item for sale on a screen and then enter an order for that item by computerized telephone line. The telecommunications industry is already a business estimated to be in the $100 billion range. *Business Week's* "Guide to Careers" estimates a growth rate of 25 percent annually for the next several years.

TYPES OF TELEPHONE JOBS

Many of the positions in telephone work are mechanic, and may not require previous education, experience, or training to be hired as an entry-level employee. Generally, if a prospective employee does not have previous training, he starts out as a frame

Fig. 8-1. Many positions in telecommunications include work in comfortable, clean locations, such as the worker here enjoys while adjusting a microwave system. COURTESY OF ANDREW CORPORATION.

wirer. In this position, he learns how to wire the frames into which many thousands of telephone lines terminate. He learns color coding, good wiring practices, how to test for good connections, and methods of locating a bad connection. In some telephone companies, a training program of up to six months may precede the actual work. Promotion into other mechanic positions is possible. Advancement may be into jobs such as central office installer, central office repairer, private-branch exchange (PBX) installer, PBX repairer, or radio repairer or mechanic. Advancement into these positions is partly on the basis of seniority, and occurs after further internal training. Study outside of work also counts toward advancement. If you have had formal study in electronics or communications or related subjects, your chances for being hired and quicker promotions are increased. Graduates of a Telecommunications Technology Associate degree course usually receive hiring preference (FIGS. 8-2 and 8-3). The students shown are enrolled in the telephone technology program of Pennsylvania State University Campus at Wilkes-Barre, PA.

Fig. 8-2. Instructor checks out oscilloscope measurements on actual telephone equipment made by student in Telephone Technology Associate Degree Program. COURTESY OF C-TEC CORPORATION.

Fig. 8-3. Students in telephone technology course receive instruction on telephone electronic equipment installed in cabinets. COURTESY OF C-TEC CORPORATION.

According to the U.S. Department of Labor, there are about 73,000 of these mechanics jobs. Pay scales range from $6 to $16 per hour. The middle 50 percent of workers earn between $8.90 and $14.20 per hour. Membership in a union may be a requirement in some companies after a 90-day probationary period.

EMPLOYMENT AND WAGES

A recent study of the telecommunications field by Pennsylvania State University showed that 74 percent of graduating students holding the associate degree in telecommunications obtained jobs in that field. Of the remainder, 9 percent continued additional study and some of the balance selected other positions of their choice in related work. The study found that the average entry-level salary was $19,059 with advancement on the job resulting in increases in wages.

To get started in the telecommunications field it is not essential to have an associate or higher degree. Completion of a technical or trade school course or a home-study course is usually acceptable. The course should include basic mathematics, electricity, and electronics, plus some training or study in dealing with customers and the public.

WORKING CONDITIONS

Because communications is so essential to modern business and society, communications technicians, engineers, and other workers are provided with the best test instruments and tools. Also, in many jobs, the work is indoors under pleasant conditions, in clean well-lighted surroundings. In other situations, the work is at customers' premises, and sometimes out of doors, as when installing a satellite transmitter or receiver, when working on vehicles equipped with mobile telephones, or at a repeater station. Some jobs require standing for long periods, working on a ladder, lifting, or carrying equipment. Night and weekend work on the basis of seniority or on a rotating shift is not unusual. In some work a telephone headset, a safety helmet, and/or a tool belt must be worn. Benefits usually include full or partially-paid medical and hospitalization, pension plan, and vacation.

MODERN TELECOMMUNICATIONS SYSTEMS

Some of the modern equipment used by telephone companies and others engaged in telecommunications work is now quite sophisticated compared to older equipment. For a technician to work on much of this equipment requires training in electricity and electronics. Technicians frequently work on digital systems, computers, microprocessors, electronic switching, and teleconferencing, video, fiberoptic, microwave, and satellite systems (FIG. 8-4). In this work, employers usually prefer someone with at least an Associate degree or the equivalent in electronics technology or electronic communications.

TWO-WAY RADIO WORK

Telecommunications companies and other organizations employ technicians for the installation, adjustment, troubleshooting, testing, and repair of two-way radio equipment. This equipment includes transmitters and receivers at each of two or more fixed points, or between fixed points and vehicles, or between vehicles. Some industries that use two-way radio are trucking companies, plumbing contractors, construction organizations, protection/security companies, electrical contractors, auto wreckers or towers, radio/TV stations, landscapers, utilities, and lumber/building suppliers. Other users of two-way radio include public safety organizations such as police and fire departments, railroads, pipeline companies, and airlines. Generally, these organizations have their equipment installed and repaired by independent or third-party maintenance organizations specializing in this type of work. Technicians employed for two-way radio work often have an associate degree in electronics or electronic communications. Courses should include substantial study of radio frequency fundamentals, transmitters, receivers, and the measurement and testing of these systems.

To locate companies who employ technicians in two-way radio, look in the Yellow Pages of your local telephone directory under Radio Communications Equipment and Systems.

Fig. 8-4. Some telephone company workers install and repair microwave radio systems. COURTESY OF ANDREW CORPORATION.

The U.S. Government and governments of other countries use two-way and other radio systems and employ technicians to troubleshoot and repair them. State, county, and city governments also employ two-way radio equipment, and some of these hire their own technician or technicians to take care of their equipment. Others have their equipment taken care of by independent or third-party service organizations. Some technicians

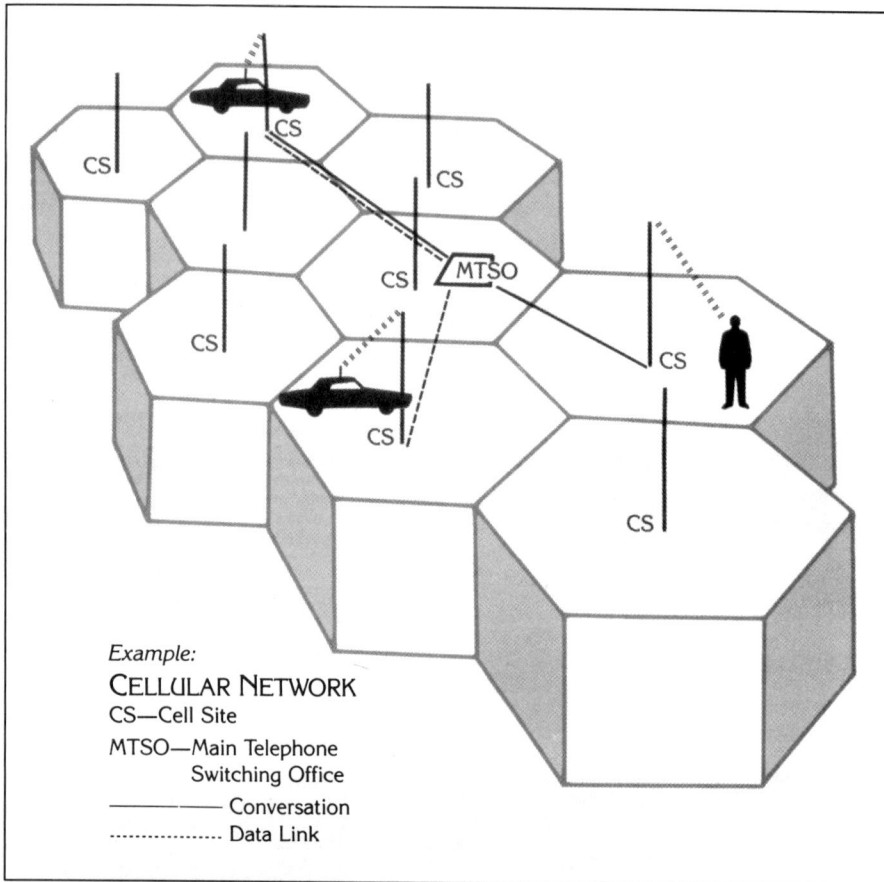

Example:
CELLULAR NETWORK
CS—Cell Site
MTSO—Main Telephone
      Switching Office
———————— Conversation
··················· Data Link

Fig. 8-5. New cellular mobile telephone technology produces higher-quality transmission and greatly improved channel availability over conventional mobile telephone service. Cellular service operates with a number of low-powered transmitters located in adjacent sites. As the user moves from one site to another, the signal is passed via microprocessors to the adjoining transmitter. The signal, or mobile conversation, maintains constant strength even as the user moves away from the transmitter. COURTESY OF C-TEC CORPORATION.

specialize in repair of two-way radio equipment and work on a self-employed basis.

CELLULAR RADIO TELEPHONE

Cellular radio (FIG. 8-5) is a form of two way radio that came on the scene in the early 1980s. Cellular radio involves the installation of a special two-way radio system in one or more vehicles. The mobile cellular radio includes a telephone handset. The operator of the auto or truck can use his cellular radio to dial a personal or business telephone, or a business or personal telephone can dial the number assigned to the cellular radio telephone. Cellular radio (cellular telephone) systems companies are listed under Radio Communications Equipment and Systems in the Yellow Pages of telephone directories of larger cities. Cellular radio is a fast-growing segment of electronics and opportunities exist to get in on the ground floor as a technician in this field.

❖ 9

Computer Servicing and Troubleshooting

T HE FIELD OF COMPUTER TROUBLESHOOTING AND REPAIR IS RAPIDLY growing, rewarding, and worth considering if you're thinking about electronics as a career. One of the most recent surveys indicated that there are upwards of 50,000 persons employed as service technicians. While many are employed by manufacturers and local or regional wholesalers, a number also are employed by companies, the military, and governments using significant numbers of computers in their daily activities.

The field of computer servicing and troubleshooting continues to grow rapidly. The best opportunities for employment as a computer service technician exist in metropolitan areas; but more and more independently owned computer repair businesses are appearing also in smaller cities, due partly to the number of personal computers purchased by individuals. Some of these smaller shops are factory-authorized repair stations for one or more manufacturers and handle warranty and after-warranty repairs for individuals and businesses.

NATURE OF COMPUTER SERVICE WORK

Troubleshooting and repair in the computer field involves not only computers, but all the related electrical and electronic equipment the computers are connected to including tape and disk equipment, printers, remote terminals, and cathode ray tube (CRT) terminals, and so on. The completely-qualified service technician may be called on to work on any or all of these. Some of the equipment may be located miles from the main computer, so extensive travel from one customer to another or the same customer or employer is not unusual. In cases where equipment cannot be repaired or put back in working condition at the customers' location, it is returned to a central repair station where either the technician who made the call, or a different technician, can carry out more extensive analysis or troubleshooting.

When failure of a computer occurs, the technician is expected to quickly find and cure the trouble. Many technicians are "on call"; that is, they are expected to be available on a 24-hour basis, subject to call, day or night, for a large or important computer system. The repair of a computer doesn't necessarily require the searching out of a particular small part among thousands of other parts. It usually involves replacing a circuit board or module to get the system operating quickly as possible once the board or module containing the source of the trouble is located. Sometimes the board or module is simply discarded or, as time permits back at the central shop, the individual defective part is located and replaced (FIG. 9-1). The board is then checked out and used as a replacement later.

In many cases, the particular section or circuit board that is defective is identified by running a special diagnostic program designed for rapid troubleshooting.

The actual troubleshooting and repair work involves the use of typical hand tools used in electronics work, such as pliers, cutters, and soldering iron. Use of electronic test instruments, some of them quite sophisticated, is also required. These include analog and digital meters, oscilloscopes (FIG. 9-2), distortion meters, line voltage analyzers, and others (FIG. 9-3). The instruments and tools are used not only when troubleshooting but when installing new computer equipment including cables and some of the electrical wiring.

Fig. 9-1. Technician operating capacitor-inductor analyzer. COURTESY OF SEN-
CORE ELECTRONIC TEST INSTRUMENTS.

Computer service technicians must carry on their work in a pleasant and professional manner since they are dealing with customers who directly or indirectly are paying their wages. They must offer advice without being offensive to customers on how to keep their equipment working properly. They may train customers or technicians on preventive or corrective maintenance. Sometimes, a trouble call turns up nothing actually wrong with the equipment but a problem with the use of the equipment; then the technician must tactfully convince the customer of that fact.

Computer technicians often work overtime, or are asked to come in to work again for an emergency after they have gone home, but they are usually scheduled to work a 40-hour week. Work for manufacturers or for the large computer service organizations usually includes good benefits such as vacation, medical and hospitalization at least partly paid, pension plan, time off for illness or serious business, and sometimes, profit sharing.

Great physical strength is not required. (Sometimes, however, the technician must lift or help to lift relatively heavy items and test equipment.) The work is not dangerous aside from being exposed to the possibility of electrical shock or burns. In some computer technician positions, the employer provides a company car for calling on customers. In other cases, the technician is paid on a per-miles-travelled basis. If work is away from home, especially overnight, meals and lodging are also company-paid or reimbursed.

Fig. 9-2. Knowing how to use a computer in troubleshooting computers is important. COURTESY OF TEKTRONIX, INC.

WAGE RATES

Beginners or trainees as computer service technicians started out at about $270 per week in 1984, according to the Occupational Outlook Handbook, 1986–87, published by the U.S. Department of Labor. The median weekly earnings at that time were about $480, with the middle 50 percent earning between $375 and $625, and the top 10 percent earning $740 a week or more. Actual wages for an individual technician depend on the length of experience; his skill, training, and type of work; and the employer.

EDUCATION AND TRAINING

To get a start into the field of computer troubleshooting and electronics, most employers look for prospects who have had at least one to two years of specialized education or training in basic mathematics, electricity, electronics, and data processing, including digital principles and basic logic circuits. The training

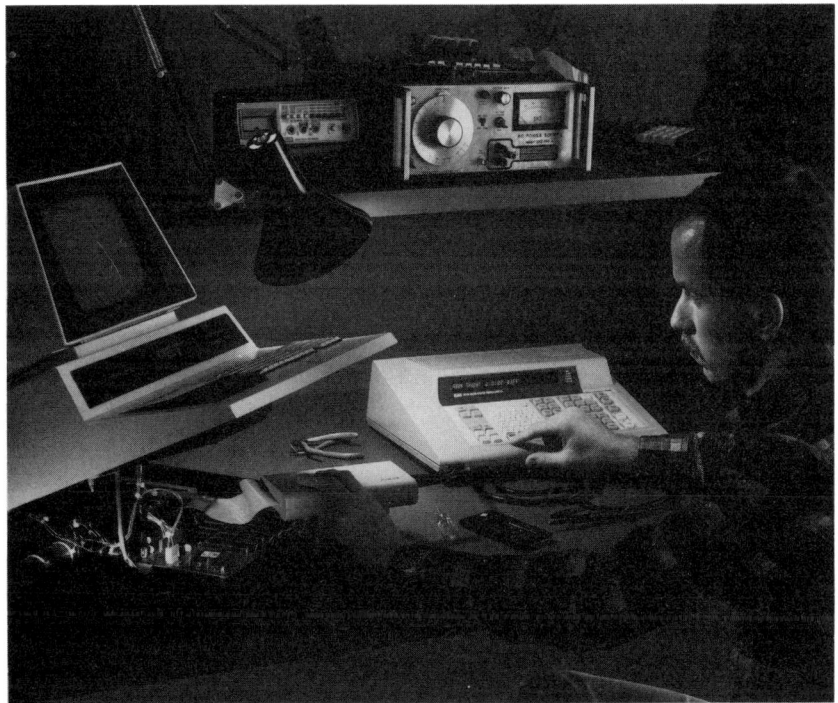

Fig. 9-3. Technician here is using a Micro-System Troubleshooter for servicing a personal computer. Special-purpose instruments like this speed up fault finding. COURTESY OF JOHN FLUKE MFG. CO.

may be from public or vocational technical schools, private trade schools, junior colleges, correspondence courses, schools offering associate degrees in electronics, or colleges. Armed forces training and experience in other related fields such as TV repair and business-machine repair are also sometimes acceptable. Some study of math, science, and physics is helpful.

A technician who has had little experience is usually put into a training program which may last from one to six months. This training may include a review of math and electronics fundamentals plus practical hands-on training on the company's products. Sometimes, after a period of actual work on one product, the technician is brought back into the central office or factory for more advanced training or training on another product. Technicians must sometimes also take updating or upgrading courses when products are replaced or modified or new products are developed. The technician should have some ability in programming of the equipment he services. This programming knowledge helps in setting up situations in which errors are said to have occurred.

ADVANCEMENT

Some thoroughly trained and experienced technicians are assigned to engineers to help in the design of new equipment or the modification of existing systems. Promotion to supervisor or service manager is one of the advancement paths available to qualified technicians. Other advancement routes include becoming programmers, sales engineers, special customer service engineers, and installation planners.

GROWTH OF FIELD

The need by the computer industry for computer technicians is expected to increase through the middle 1990s because of the increasing computer equipment being installed in industries and other organizations of all sizes. However, as the field develops, better ways of troubleshooting are also being developed. Therefore, while the field expands, the need for technicians also increases, but at a slightly slower rate.

Technical Writing and Instructing

A N AREA OF ELECTRONICS THAT PROVIDES A GOOD LIVING FOR SOME people and at least some extra income for many others is writing articles, lessons, and books. To be a writer, you needn't necessarily have a perfect mastery of the English language, or be a perfect speller, or use flawless grammar. Publishers often have editors to polish up things like that. In electronics it's more important to be able to put down on paper the fruits of your knowledge and experience.

MAGAZINE ARTICLES

Electronics publications are always on the lookout for writers of interesting articles. Some of them pay reasonably well for such articles. Some of them will provide a list of subjects on which they would like to have articles. Electronics publications might be in the market for articles on compact disk (CD) players, home TV satellite reception, AM stereo, stereo TV, VCR and camcorder troubleshooting and repair, projection TV, computers, construction and experimenter articles, new TV circuits, audio hi-fi and stereo, and others. Magazines regularly also publish articles on electronics fundamentals such as filters, transistor principles, regulated power supplies, and logic circuits.

After you've gained some experience in electronics, the best way to find out what magazines are looking for is to drop a note to the attention of the editor saying you're interested in writing and asking what kind of articles he is looking for. You might have some ideas of your own which you could suggest. If he's interested, he'll probably ask for a brief outline, or a few sample paragraphs. If it turns out that the editor likes your work, it's possible to negotiate, in a friendly way, for a better fee. And if you're really good, you could be a regular contributing author, or write a regular column. Check with the magazines listed in Appendix B of this book.

WRITING BOOKS AND LESSONS

Other writers in electronics make a living or supplement their income by writing books. Book publishers are always in the market for good electronics writers. (The publisher of this book is one of them; others are included in the list of books publishers provided in Appendix C.) Books that are in demand cover the same general subject areas as magazine articles. There are also handbooks, books on experiments, "cookbooks" on troubleshooting, and books on every conceivable aspect of computers.

Before you write your book, the publisher will want an outline in some detail. After approval of the outline, you'll discuss a date for delivery of the completed manuscript and the fee you'll receive. You might be paid some modest advance to cover some of your incidental costs while you write the book. Upon completion, if it's accepted, your previous agreement with the publisher determines whether you're paid another advance, or paid fully for selling the manuscript and all rights to the publisher, or whether you wait until the book starts selling and then you're paid royalties (a certain percentage of the selling price of the book). The more books sold, the bigger your monthly or quarterly check. If you're lucky and the book sells well, you'll probably receive checks for three to five years and then it will be time to update the book. You will probably be offered the job of doing the updating. If the book doesn't sell well, it will probably be dropped after the first printing. This needn't mean you're a poor author—it might only be evidence that you and the publisher picked a topic that didn't go over so well.

Writing electronics lessons is another source of additional income. Write to the directors of home study, or correspondence, schools, or the director of research and development, mentioning that you are interested in writing or updating lessons for them. Describe briefly some of the writing you've already done and what your main field of expertise is. Correspondence schools don't look for authors on the very latest developments in electronics. They've learned that sometimes by the time a lesson on a new development gets into print, it has come and gone. Some things don't prove to be as practical or popular as at first they seem. Instead, correspondence schools look for authors having a solid foundation in the fundamentals and the ability to organize their material clearly, accurately, and logically. The ability to keep the home-study student's interest is essential.

If the school thinks you'll make a good author for them, they will either provide the subject area and an outline and ask for a sample paragraph or two, or they will ask you to submit the outline with a sample paragraph. Lessons range in size from 30 to 100 printed pages and from 20 to 50 illustrations. Illustrations should be neat and legible hand sketches (they'll be done over professionally by the school's illustrators), or glossy photographs 5″ × 7″ or larger, or illustrations provided by manufacturers or other sources from whom you've received clearance. When printed, your name will probably go on the cover of the book as the author.

Schools that are members of the National Home Study Council are good prospects.

TECHNICAL WRITING

Many industrial companies as well as government and military organizations employ technical writers. Many women have been attracted to this field, as technical writers are generally well paid and there are good opportunities for advancement. Many authors and tech writers, whether self-employed or working for someone else, find that using a word processor (FIG. 10-1) greatly speeds up their work and makes corrections and changes very convenient.

Fig. 10-1. Word processors have made the work of authors and technical writers faster and easier. COURTESY OF BILL RISSE.

An electronics technical writer usually works closely with engineers who design equipment. The writer's job is often to prepare an operating manual and instruction on maintaining and trouble-shooting the equipment (FIG. 10-2). There is usually a section on theory of operation, describing how the circuits work. Engineers don't usually have the time or inclination to write these manuals, so it's up to someone else with good electronics knowledge to get the information from the engineers and put it into words that will be helpful to the customer purchasing the company's products. Good operating and maintenance manuals help sell the products. Users of electronics equipment soon get to know which manufacturers have good operating and maintenance manuals and which ones do not.

Some technical writers must adhere to military specifications (mil specs) when organizing and writing manuals; this is learned on the job in most cases. If you have a good electronics foundation, many companies will provide the specialized training for you to become familiar with the needs of the military.

Jobs for technical writers are often listed in the Sunday editions of the larger newspapers. Sometimes they are listed under technical or engineering employment agencies.

INDUSTRIAL AND TRADE SCHOOL INSTRUCTOR

Industrial training departments of some of the larger companies frequently have openings for electronics instructors. Usually they look for people with good electronics training themselves, and then provide product training before assigning the instructors to classroom or seminar work. In a position of this type, you'll either be teaching employees or customers the technical aspects of your company's products or equipment. Usually, the students are sent to your plant and you teach them in a company

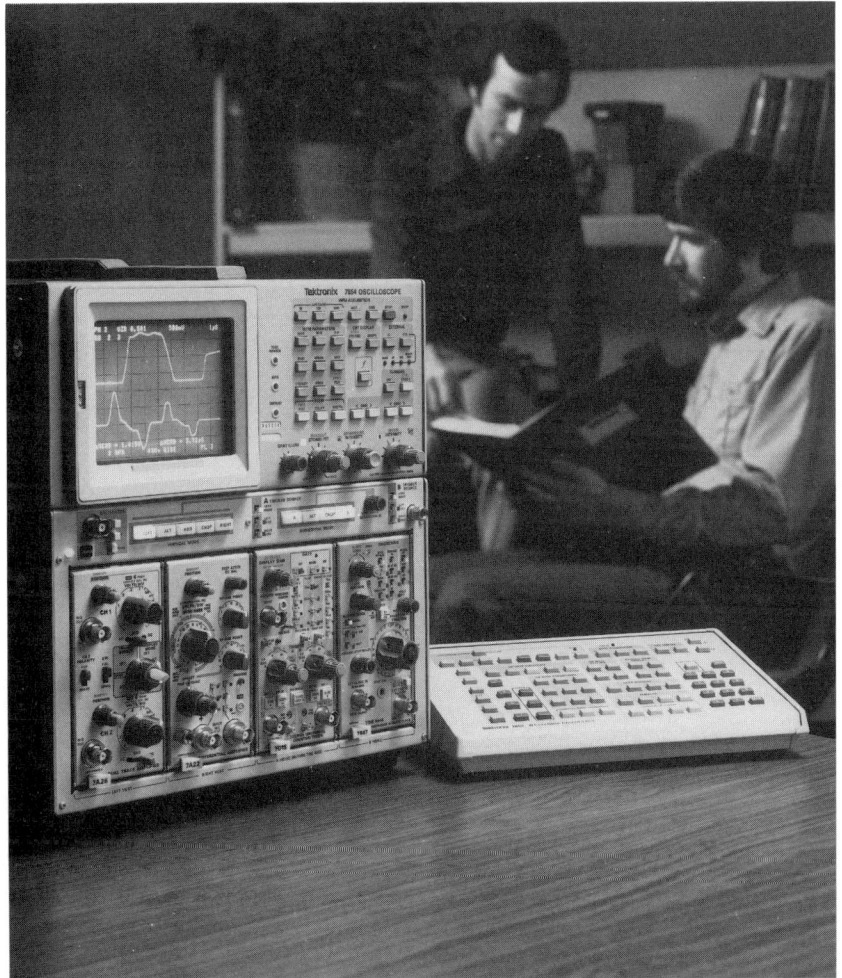

Fig. 10-2. A tech writer and an engineer discuss preparation of an operating manual for new equipment. COURTESY OF TEKTRONIX, INC.

classroom. In other cases, you are sent to a client's premises or to a central location to present a seminar or class on product training.

Vocational, trade, junior colleges, and technical institutes also hire instructors with a good foundation in electronics (FIG. 10-3). In some of these positions, the main requirement is good knowledge of the course you'll be teaching and the ability to interface with students and impart your knowledge to them. A degree in electronics is often not essential when hired, but, if hired full-time, you might be expected to study toward obtaining a degree including taking some teacher education courses.

Generally, instructor positions pay reasonably well and include advancement and good benefits including health insurance and hospitalization, vacation, pension, and paid holidays.

ADVANCEMENT POSSIBILITIES IN WRITING AND INSTRUCTING

Work as a writer or instructor offers excellent advancement opportunities into positions of greater responsibility. The ability

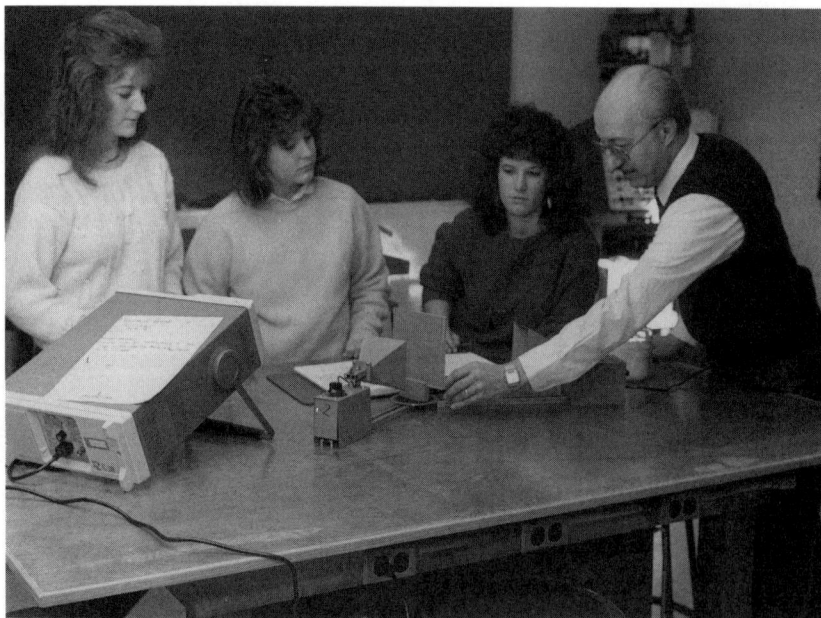

Fig. 10-3. The instructor is demonstrating theory and principles of micro-wave transmission to students enrolled in telecommunications associate degree course. COURTESY OF C-TEC CORPORATION.

to write, and to get your ideas across to others, leads to higher-level jobs in management, marketing and public relations, education, training, and research.

An organization that is dedicated to the advancement of its members in technical communications is the Society for Technical Communications (STC). Technical communications includes writers and editors, technical instructors, illustrators, and printing and publishing personnel. Members are kept aware of job and position openings of all types, and are provided information valuable for advancement in their work. There are local STC chapters in most medium and large cities in the U.S. and Canada. Nonmembers are usually welcome to meetings. (STC's address can be found in Appendix G.) Several grades of membership are available for the beginner and for the professional.

❖ 11

Education and Training for Electronics

T HERE ARE A NUMBER OF WAYS TO GET THE EDUCATION AND TRAIN-
ing needed to work in electronics, including enrolling in a
four- or five-year college course in electrical or electronics engineering; signing up for a two or three year study program in
electronics technology or electronics engineering technology; attending a local electronics program at a junior college, community college, technical institute or college extension (FIG. 11-1);
studying electronics at an adult vocational technical school; taking night courses in electronics for adults; enrolling in an accredited correspondence course, or learning electronics through on-
the-job training, an apprenticeship program, military service, or
through self-teaching.

The path you choose to learn electronics depends on your cir-
cumstances and how much time you have available.

COLLEGE PROGRAMS

Many colleges and universities offer excellent electronics programs. To get a list of those in your state, write to the Department of Education at the state capital. Many colleges cover electronics in their electrical engineering programs. If you complete all the prerequisites for a degree in electronics, your degree will probably be either in Electronics Engineering or in Electrical Engineering with a major in electronics. Some college programs in electrical and electronics engineering emphasize industrial electronics, some communications, some computers; others concentrate on power applications or power generation and distribution. Some colleges also offer electronics under their physics department — these electronics programs are slightly more science-oriented.

To be admitted into an electronics engineering program in college, it's usually necessary to have completed the academic or science high-school curriculum. This means that you need some high school science, physics, math, and chemistry courses. If you're missing some of this, all is not lost, however; colleges and local high schools offer remedial courses and other study programs.

Fig. 11-1. Technical institute instructor demonstrating lab equipment to students. COURTESY OF JOHNSON TECHNICAL INSTITUTE.

In college you'll have courses that are heavy in math, science, and electronics theory, plus plenty of lab courses to give you practical experience and knowledge. In the lab classes, you'll probably work with one to four other students (FIG. 11-2). Lab classes should give you knowledge in circuit building, the ability to analyze circuits, and skill in writing up reports. All three are essential to advancing in your career later. In lab classes, don't overspecialize, such as always being the one who takes the notes and writes up the reports, or the one who knows the theory but can't put in on paper intelligently, or the one who always builds the circuits but doesn't really know what's happening.

Many college programs now include the use of a personal computer; some have on-campus or centralized main frame computers. Some allow you to use your own computer. Often, a short non-credit course in computer programming using the BASIC language is made available to science and engineering students. Most colleges now require engineering and science students to learn more advanced languages such as FORTRAN, COBOL, and APL as part of the curriculum.

Some colleges or college engineering curriculums are not ac-

Fig. 11-2. In a college lab class, students work in small groups of two or more. COURTESY OF C-TEC CORPORATION.

credited. They may be good courses and curriculums, but they might not really qualify for transfer credit to another college, or be applicable towards a more advanced degree. Before you enroll in a program, ask the college registrar or department head if all the courses are transferable and if the program you're enrolling in is accredited by the state's department of education and by the regional accrediting association. One way to determine whether a college program in electronics or electrical engineering is accredited is by checking with the Institute of Electrical and Electronics Engineers (IEEE) or the Engineer's Council for Professional Development (ECPD). (Addresses are in Appendix G.) You can also check in a book available at most public libraries, *American Universities and Colleges*, published by the American Council on Education. (Address is in Appendix A.)

To supplement the information on electronic engineering opportunities you've learned in this book, ask at your library for *Opportunities in Electronics and Electrical Engineering*, published by VGM Career Horizons, National Textbook Company. (You can also contact the publisher directly; see Appendix A for the address.)

One more bit of advice before enrolling in a four or five year electronics engineering program. Some college courses include the word "technology" in the title of the program; for example, Bachelor of Electronic Engineering Technology, rather than Bachelor of Science in Electronics Engineering. Some employers do not have quite as high a regard for graduates of bachelor of engineering (BET) technology programs when looking for engineers as they do for bachelor of science (BS) programs. The reason for this is that they feel that the BS courses are a little higher level, going deeper into theory, than the BET programs. BET graduates generally have a better practical training, however, and in many cases can easily hold their own with the BS graduates, often excelling them. Some employers prefer BET graduates in certain positions. Both are good degrees — just be aware of these possibilities and attitudes.

COMMUNITY COLLEGES AND JUNIOR COLLEGES

Many community colleges and junior colleges offer excellent programs in electronics called Electronics Technology or Electronics Engineering Technology. They are generally two- or three-year programs, and the graduate is typically awarded the Associate in Applied Science (AAS) degree. In nearly all cases, the programs are designed as terminal programs, meaning that the main objective of the program is to prepare the graduate to enter the job market with knowledge of both theory and practical application and not necessarily to transfer to a four-year college upon graduation (FIG. 11-3). However, in many cases, transfer of some or many of the credits is possible. Many people enroll in community and junior colleges fully intending to continue their education at another institution later. Some colleges have local extensions offering two-year terminal programs. The courses in some of these may be directly transferable to a continuing BS or BET program. Be sure you ask the college you plan to enroll in later, if that's your intention. Some colleges accept such credit only into their BET programs; others accept credit in either one. It usually depends on the two-year college you've graduated from and on your grades, and sometimes it may depend on how badly the college you want to enroll in needs students.

To find out about community and junior colleges in your geographical area, contact the department of education at your state's capital.

Fig. 11-3. Junior colleges, community colleges, and technical institutes combine theory and practical applications in their two-year associate degree programs. COURTESY OF JOHNSON TECHNICAL INSTITUTE.

TECHNICAL INSTITUTES AND VO TECH SCHOOLS

Technical institute and post-secondary vocational-technical (vo-tech) schools are generally proprietary privately owned schools. Most of them are very good schools. There are around 8000 of them in the United States. Generally, they offer a two-year program of studies. Many of them award the AST (Associate in Specialized Technology) degree to graduates. Graduates of technical institutes, especially of the schools that have earned an excellent reputation in their years of operation, often have local employers standing in line to offer them jobs upon completion. Students who enroll in these schools typically do not have continuing study later at a higher-level school in mind when they enroll, but many of them do decide later that they'd like to continue their education. If you are interested, ask before you enroll what percentage of graduates have gone on to four-year colleges, and which colleges these are.

Usually technical institute and vo tech schools are located in areas where their graduates can find employment.

Some of the electronics craft or mechanics programs, or special-purpose training programs are shorter in length. Since the quality and objectives of these programs can vary considerably, the investigate them thoroughly before enrolling. Check on the reputation of the school, its employment record for graduates of its electronics programs, and accreditation. Don't overlook transferability if there's the slightest possibility you might want to continue your studies upon graduation.

Most technical institutes and vo tech schools (not including vo tech high schools) are members of the National Association of Trade and Technical Schools (NATTS). The NATTS reviews programs of its member schools on a regular schedule, and accredits the schools if the programs are approved. Ask, or check in the school's catalog, to see if it's accredited. You can obtain a catalog listing of all NATTS member schools in the U.S.: by writing the address found in Appendix A.

NIGHT SCHOOL AND CORRESPONDENCE COURSES

Perhaps you're already working and cannot attend school full-time to learn electronics. Maybe you have a family or other responsibilities. Many people get their education and obtain good jobs by either going to school at night, after work, or studying a correspondence course. Many colleges, junior colleges, community colleges, and technical institutes have night courses especially for those who cannot attend during the day or full-time. Look these up under "Schools" in the Yellow Pages in your telephone book. Some of these lead to two-year or four-year degrees. Others are special courses designed to suit the needs of local industry, so the graduates of these are reasonably assured of jobs upon graduation.

Many people now enjoying excellent positions in electronics obtained some or all of their education and training through correspondence study, also called independent study or home study. If your hours are not regular, you travel, you are not located within practical commuting distance of a school or college, or your work is seasonal, consider enrolling in a correspondence course.

Some correspondence schools even offer programs in electronics which award the AST degree upon completion (FIG. 11-4). Generally these schools are accredited by the states in which they are located, and by the National Home Study Council (NHSC). Some colleges accept transfer credit for some of the courses in those self-study programs. If you are considering a correspondence course, send for the NHSC catalog to learn who accredits the school. If it's a degree course you're interested in, and you might want to think about continuing into college study later, ask about transferability of their course credits. They will undoubtedly tell you that transfer credit depends upon the school you're transferring to later. However, they should be able to give you names of some of the colleges that have accepted such credit.

A list of accredited correspondence schools offering programs in electronics and other fields may be obtained from NHSC (address in Appendix A).

Correspondence schools also offer courses in electronics and related areas that are not degree courses; instead, they award the graduate a diploma. Courses available include computer repair, communications, TV and VCR repair, industrial electronics, and broadcasting technology. Most of these include kits or other training materials for obtaining hands-on practical experience on electronic principles, circuits, and equipment. In some, you build or troubleshoot your own computer system, VCR, TV, or CB radio.

Some include test instruments such as digital voltmeter (DVM), oscilloscope, or color-bar generator.

Learning electronics through home study is not the easiest thing you might attempt, but it is effective, and can get you off to a good start in the electronics field.

One of the advantages of study by correspondence is that some of the schools offer a toll-free telephone service so you can call in for assistance when you need help in your studies. In some cases, you might have to leave word to ask someone to call you back. You will talk to an instructor knowledgeable in that subject who is skilled in clarifying things that are puzzling.

Another service offered by at least one school is immediate exam correction by telephone. To use this service, you dial the special number, plus the serial number of the exam and your

Fig. 11-4. Some correspondence courses award the associate degree on completion of studies. COURTESY OF CLEVELAND INSTITUTE OF ELECTRONICS.

student number, and then key in numbers or symbols representing which one of four or five multiple-choice answers is the one you select for each question. You learn your grade immediately. You are also told which answers are incorrect and which sections of the lesson you should study. After some additional study, you have a second chance to pass that lesson, if you fail to pass the first time.

Some of the best known correspondence schools offering programs in electronics are listed in Appendix A.

The Cleveland Institute of Electronics, Grantham College of Engineering, and ICS Center for Degree Studies offer degree courses in electronics.

CREDIT FOR LIFE EXPERIENCE

If you've taken courses that are not really accredited college courses, and have done some studying on your own, it's possible in some cases to receive college credit for such study. College credit is also possible for certain correspondence courses, for responsible work that you've done, for papers you've presented, for articles you've written, and for other activities.

Credit of this type comes under the general heading of Life Experience. A number of colleges and organizations offer the service of evaluating life experience and education; and, for those activities that are shown to be the equivalent of similar college courses, they provide a transcript recognizing this college credit. It's not really a matter of applying and getting the credit. The burden is on you to document everything, and you may have to pass a written exam. But if you want to get credit for what you know and the work you've done, and earn credit that can apply toward a college degree, it's well worth the work and effort.

One school specializing in this is Edison State College, Trenton, NJ. Edison State College is known as a "college without walls." They also offer legitimate, fully-accredited degrees if you follow their directions on completing all the requirements. Write for their catalog.

ON THE JOB TRAINING AND APPRENTICESHIPS

Many companies offer on the job training to workers who have already had some education or experience in electronics. Not many will take on the task of training a worker in electronics from scratch; but if you've already completed a course, or have a diploma or a degree, many companies provide you with additional training on their particular products or services. For example, if you have a diploma from a technical institute or a vo tech school in electronics, and an announcement on your bulletin board is asking for applicants with an education of this type, you might find yourself in a six-month employer-paid training program to become a customer service technician to install and repair x-ray equipment or other systems that are electronics-based. Many companies like to promote from within. If you know of such a company that you'd like to work for, you might want to apply for almost any lower-paying job, anticipating that after you start work you will be selected for promotion to a position in electronics.

In many companies employing electronics workers, the work is done by employees who are members of an electrical union. In these cases, the employee is admitted into an apprenticeship program that usually covers a four-year period. An increase in pay is received for each year completed, and after successful completion of the fourth year, the employee becomes a journeyman. During the four years, a specified number of hours in various electrical and electronics specialties must be completed plus a certain number of hours of study. Apprenticeship programs in electrical and electronics work in the United States are generally a result of cooperation among the U.S. Department of Labor Employment and Training Administration, the electrical union involved, and the employer. For additional information on apprenticeships in the electrical-electronics trades, see Appendix A.

MILITARY SERVICE

One path to training and experience in electronics is through military service. If you're at all interested in military service it's possible in many cases to work out a contract with the military whereby if you enlist they will guarantee you technical training in a field of your choice. Electronics is one of the fields usually available. In the military service, it's possible to be working on the latest equipment and receive training that can excel what's offered in civilian resident schools. In exchange for this guarantee, you have to agree to remain in the armed forces for a certain length of time — typically four years — and generally the training in electronics is aimed toward military applications such as radar, sonar, or fire control. However, electronic principles are universal, and most military applications have civilian or industrial equivalents.

While in military service you also will be earning credit toward college study. Further information can be obtained from a nearby Armed Forces recruiting office.

EMPLOYER-PAID EDUCATION AND TRAINING

Many companies have a policy stating that they will pay for courses you take in certain cases (FIG. 11-5). If you are lucky enough to be working for such a company, you may or may not have to obtain approval before enrolling in a course or study program. Some companies require that the course be job-related. Sometimes you are reimbursed only after completing the course successfully. In a few cases, companies reimburse based on the

Fig. 11-5. Some companies conduct their own internal training programs. Many also reimburse employees upon completion of related courses of colleges or other education or training institutions. COURTESY OF C-TEC CORPORATION.

grade earned—maybe 90 percent reimbursement for an A grade, 80 percent for a B grade, and 70 percent for a C grade. Check with your personnel department and get complete details on the policies and possibilities. It's possible to earn a degree, studying nights, while your education is being paid for by your employer. At the same time, you're advancing in your job and planning for your future.

PAYING FOR YOUR EDUCATION

Unless you're in the military service, or are working for an employer who will pay for some or all of your education, it's necessary to somehow come up with the money to pay for it. Perhaps you've saved some of the money. If so, that can take care of at least some of the costs. If you're like most people, you're not that fortunate.

Grants and low-cost loans are available to many students in full-time study programs. Check with the financial aid officer at the school where you're thinking about enrolling. Federal and state grants might be a source of money that does not have to be repaid if you're financially eligible. There are also sources of low-interest loans, and there are work-study programs. In a work-study program, you are placed in a job where you work and are paid for the work a certain number of hours per week.

To learn more about funds that are available for college and other full-time courses write to the department of education of your state.

❖ 12

Advancing in Job and Career

HOW FAR YOU CAN ADVANCE IN THE FIELD OF ELECTRONICS depends on how well you're prepared. Preparation means obtaining the best education and training possible, and also getting your own thoughts and objectives together on what you'd like to aim for. Think ahead on what you'd like to be doing five years from now. Then work toward that goal. If it does not seem, after a year to two years, that you're making progress, revise your plan. Revising your plan might mean looking for another employer, or it might mean going to night school, or taking a correspondence course to upgrade your skills. Sometimes, the simple fact that you're taking a course in your free time gets the attention of your employer, letting him know you want to advance in your work. Some correspondence schools will send a Progress Report to your employer if you request it, which also reminds him that you're a serious worker and really want to get ahead.

Sometimes advancement means becoming the foreman, supervisor, or department manager. In that case it may help for you to study a course in modern supervision or management. At least studying such a course will give you some idea whether this is the path you'd like to follow.

If you're in a job now that is not directly electronics-related, but you work for a relatively large organization, chances are that your organization needs someone trained in electronics. For example, most manufacturers are now changing over to electronic-controlled production which requires people skilled in electronics to keep the equipment operating. As an employee who has been trained in electronics, perhaps you can be assigned to help install and maintain the electronic equipment. Let your boss know that work of that type would be of interest to you. Possibly you'd be sent to the factory where the electronic controls are made for specialized training. Banks, schools, colleges, large department stores, utilities, municipal governments, and machine shops are examples of other employers that previously didn't employ electronics-trained people but do now. If you work for such an organization, perhaps you can get in on the ground floor as the company electronics technician or engineer. Later, the department may grow, and you could become the department head.

It's not unusual for people to change employers in order to advance in a career. In your present job, there may be nowhere for you to advance, or perhaps you're being overlooked. If such is the case, stay where you are while you look for another position. If you quit, it's harder to find another job while you're out of work. Employers prefer to hire someone who's working rather than someone who is unemployed. This is unfortunate, but it's a fact of life.

However, if there are opportunities within your own company, think it over seriously before making up your mind to leave. You may be giving up seniority or pension or profit-sharing benefits by going to work for someone else.

OPPORTUNITIES FOR ADVANCEMENT

As a general rule, after you have gained experience or on-the-job training, or a combination of these, you should see your pay increasing steadily. Also, you should be in line for a promotion. Your chances for advancement become even greater if you've used some of your spare time for additional study of some kind

108

(electronics courses, courses in management or supervision, or business courses).

However, as in any other field, advancement is not determined completely on a person's job performance and education. Other factors that have a bearing on your progress include personal appearance, character, personality, and your ability to keep your employer's interests foremost.

Advancement opportunities within a company come about either because new jobs have been created due to company growth, or due to the need to replace persons who have left for other work, retirement, or any of a number of other reasons. Many companies like to use people within the organization to fill positions that open up. Such a policy tends to enhance employee morale; they can see that there are really chances for advancement. Promotion from within means that the employee being promoted is already familiar with the company's policies, procedures, and methods. Improved company loyalty is another benefit. Also, a shorter training program is needed.

ADVANCEMENT QUALIFICATIONS

Your chances for advancement improve if you have certain essential qualifications: good work performance, job knowledge, dependability, education and training, good character and personality, and a proper attitude toward your job and your employer.

How well you do your job is another name for work performance. It shows what you have done in the past and it reflects your capabilities and your attitude. Work performance can also be an indicator of your potential for greater responsibility.

Knowledge of your job is obtained through experience and on-the-job training, either in your present position or in previous positions with the same employer or previous employers. Job knowledge is also obtained through study of related courses on your own or through attending or participating in employer-sponsored programs (FIG. 12-1).

Personality is nearly always an important factor in job advancement. It's possible to be serious in your work and yet have a pleasing personality and an eagerness to learn. Also, neatness in your personal appearance can also determine whether you'll be

Fig. 12-1. Keeping up with developments either through employer-sponsored programs or enrollment in appropriate courses helps in career advancement. COURTESY OF C-TEC CORPORATION.

considered seriously for greater responsibility. Honesty and reliability are also important.

Qualities of leadership and supervision are important if the advancement is into a management or supervisory position. These qualities can be learned or improved by on-the-job seminars on modern supervision or through study on your own.

ADVANCEMENT PATHS

Advancement paths can take any of several different directions. The most common advancement is within the same job or position to more responsible assignments. An example is advancement from routine testing and troubleshooting work to a position performing complex tests or troubleshooting difficult problems.

Advancement from one job classification to another also helps. For example, you might be promoted from electronics production-line tester to tracing down hard-to-find bugs at the end of the line (FIG. 12-2). Another example is promotion from engineering

Fig. 12-2. The technician here is using an oscilloscope, with a number of other instruments also at his disposal. COURTESY OF SENCORE ELECTRONICS TEST INSTRUMENTS.

aide to technician, or from technician to assistant engineer, or from engineer to engineering manager. In those kinds of advancement, the job class changes along with the kinds of duties performed.

There are also advancements to supervisory responsibilities. A worker might go from technician to group leader, chief technician, foreman, or manager. Titles of this type vary from one organization to another. Generally speaking, whenever there is a group of hourly-paid workers, a supervisor is assigned to represent management and to be responsible for the overall quantity and quality of the work performed.

Advancement may also be in the form of transfer to another department or to another type of work. For example, an electronics technician might be transferred or reclassified from technician to purchasing agent, sales technician, or customer service technician. He might be asked to help train new workers, or be an instructor training customer technicians, or to move into a technical writing position preparing operating and maintenance manuals on the company's electronic products.

Even though an advancement does not necessarily mean more money immediately, it may give you the opportunity to broaden your qualifications. The knowledge and experience you gain may lead to future opportunities and significantly greater financial rewards.

Actually, advancement possibilities are almost limitless. Many an electronics technician and engineer has advanced to the top of his organization.

ADVANCEMENT IN CONSUMER ELECTRONICS SERVICING

Technicians who service television, radio, VCRs, stereos, and other equipment owned by the general public are often referred to as customer electronic service technicians (FIG. 12-3). These technicians who work in large repair shops have the opportunity for promotion to foreman or service manager. Some technicians work for companies that manufacture consumer electronic equipment. These workers can progress to higher paying positions: parts manager, sales engineer, service training instructor, or publications manager. A technician employed in a one- or two-man shop may have the opportunity to become a partner with his employer, or to become the manager of a new shop if the owner decides to expand. In an arrangement of this type, he advances as his employer's business grows.

ADVANCEMENT IN ELECTRONICS ENGINEERING

As an electronics engineer, you have two main advancement paths available to you: up through more responsible engineering positions, and through management positions. In electronics, there is a shortage of engineers with good management potential. Many engineers are not interested in the management or business side of electronics. If you are planning to study electronics engineering in college, be sure to include courses that will polish your ability to communicate with others. Also, if your schedule permits, include at least one course on business or management. If you cannot fit these in your college schedule, don't overlook the possibility of taking business and management courses after you've graduated.

The engineering department of a manufacturer generally includes engineering assistants and laboratory technicians, engineering aids, mathematical specialists, electronics technicians,

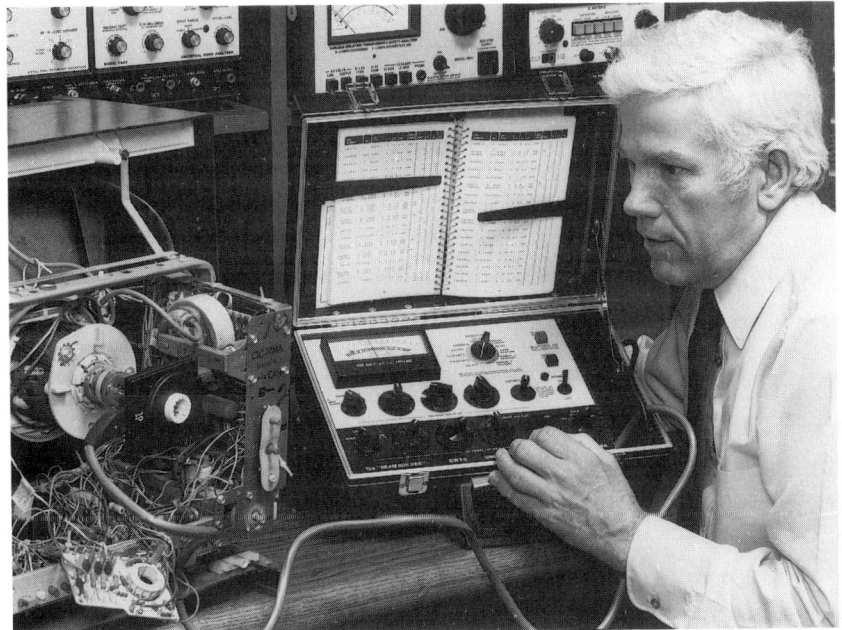

Fig. 12-3. Technician using TV CRT checker. COURTESY OF SENCORE ELECTRONIC TEST INSTRUMENTS.

and drafters and technical illustrators. Usually, an engineer heads a department of this type. There may be additional engineers, such as design engineers, sales engineers, or others.

The military and space industries offer good positions with advancement opportunities for engineers. So does the computer industry. And manufacturers of integrated circuits employ a large number of engineers. Other industries employing electronics engineers include the automobile manufacturing industry, aircraft manufacturers, the medical equipment field, and the telecommunications industry.

ADVANCEMENT IN SALES AND ADVERTISING

The selling and advertising of electronics products require knowledgeable people to describe the advantages and specifications of those products. Technical people are needed to guide sales personnel and those who write advertising materials so that sales approaches and advertising brochures sound professional and are accurate. Promotions to sales and marketing positions may be available to many technicians and engineers in electronics work. A qualified person who goes into sales or advertising may find further advancement possible, to sales manager, service representative, or director of promotional publications.

ADVANCEMENT THROUGH ASSOCIATION MEMBERSHIP

Some of the advantages of joining an association are discussed in later parts of this book; let's briefly discuss job advancement through association membership.

One of the advantages is that help is available with certification and licensing. Becoming certified as an electronics technician means that you can use the letters CET (certified electronics technician) after your name. If you're a broadcast technician, you can become certified by the Society of Broadcast Engineers which provides additional testimony to your experience and qualifications.

An association can also provide you with helpful information about becoming licensed. Technicians in the communications industry may find it helpful or necessary to acquire an FCC License. An electronics engineer may be interested in becoming a Professional Engineer (PE); this requires a certain number of years of experience plus passing an examination provided by the department of the state that is responsible for licensing. Association membership can be helpful to technicians and engineers toward attaining these goals and toward position advancement.

An association also is a source of information on where to obtain courses, training, film, tapes, and books in various topics. Association membership also provides the opportunity to contact other persons doing the same type of work you are doing. As a result of such contact, ideas for solving problems often result. Also, members are generally advised of various job openings. Members of associations frequently are the head technicians, supervisors, engineers, and managers of their companies and it is not unusual for them to be on the lookout at association meetings for likely technicians or engineers to hire.

An association offers you strength in numbers—for example, let's say a certain law is up for consideration that will not be to the benefit of your kind of work. Your association can generate a petition and letter to present to legislators.

Membership in an association provides you with the opportunity to attend national and regional conventions and meetings. At such gatherings, you can examine the newest products and attend seminars or meetings on the latest technical developments. You can also meet manufacturers representatives and learn of ideas to help you in your work.

See Appendix G for a list of professional associations.

❖ 13

Sources of Technical Literature

O NE IMPORTANT WAY TO KEEP UP WITH DEVELOPMENTS IN THE field of electronics is to read the technical publications that cover your particular area of electronics.

MAGAZINES

Magazines are one type of publication that are available either by subscription or at the magazine racks of the larger newstands and drug and department stores. If you are a student, you should not spend an excessive amount of time reading magazines — you will want to devote as much time as possible to your studies — but some limited magazine reading will give you a feel for the present practical developments. Also, some magazines include tutorial articles; that is, articles that cover general principles and theory. These may be directly helpful in your studies.

If you're already working in electronics, unless you somehow keep up with developments, you'll slowly slip behind. Within a matter of five years or so you'll be on the way to becoming obsolete except, perhaps, in your own narrow field of specialization. Reading magazines and other technical publications is important to your advancing career in electronics. Appendix B lists many of the more important magazines that are presently available. Of interest to service technicians are magazines such as *Radio Electronics*, *Appliance Service News*, *Computer/Electronic Service News*, *Electronic Technician's News*, *Electronic Service News*, and *Professional Electronics*. If you're still a student, you might want to read magazines that cover subjects on a more popular or lighter level, such as *Personal Electronics*, *Popular Computing*, or even *Radio Electronics*. (Not all the articles in the latter publication are suited for beginning students.) If you're a service dealer — that is, if you sell TVs, computers, and VCRs as well as service them — you'll be interested in magazines such as *BYTE*, *Popular Computing*, *Dealerscope*, *Electronic Technician Dealer*, *Satellite Retailer*, *S&VC*, or *Two-Way Radio Dealer*.

Specialized magazines include *Business Radio*, *CEE* (Electrical Construction), *Communications*, *CB Radio Times*, *Electronic Test*, *Industrial Research and Development*, *Instrumentation and Control News*, *Medical Electronics*, *Microwave System News*, *Pro Sound News*, and *Robotics Age*.

If you're in broadcasting or in related activities, you'll want to consider subscribing to *Broadcast Engineering*, *Microwave System News*, *Radio World*, *SBE Signal*, *SATVISION*, *SMPTE Journal*, *Television Broadcast Communications*, or *Video Systems*. Some magazines, on a higher technical level which might be of interest mainly to engineers, include *Electronic Design*, *Computer Design*, *Digital Design*, and *Electronics*. *Telephony* covers the field of telephone communications.

If you're not able to find some of these locally at newsstands, public or school libraries, or on the job, and you're not sure whether you want to subscribe, write to the publisher and ask for a sample copy. Most publishers are happy to provide a copy.

Some of the magazines listed are available free to technicians, dealers, and engineers who are working in the field they cover. When you're asking for a sample copy, ask whether complementary subscriptions are available, and state what your work is. Some of the magazines have a mail-in card enclosed for free subscriptions to those qualified. Others magazines are available only to those working in that field, so the publisher can better assure advertisers that they'll have a highly-interested readership who are potential purchasers of their products.

Reduced rates are often available for subscription to some of the more popular magazines. You can find these reduced-rate applications at school and college libraries. They are intended mainly for students, but you may get some offers also at home if you are on mailing lists. If you are studying a correspondence course, your school should have applications for reduced-rate subscriptions available. Some reduced-rates offer subscriptions at one-half the *cover* price of the magazine. Others offer even cheaper subscriptions: these might be close to one-half the regular *subscription* price. Lists of other literature are regularly available to members of associations.

MANUFACTURERS' LITERATURE

If you are interested in certain electronic components or systems and cannot obtain them through your school or on the job, try the local distributor. In many cases, they have tear sheets or brochures describing their capacitors, test instruments, CB radios, and so on. If you have no luck there, call or write directly to the manufacturer. Manufacturers often have ads in the magazines you'll be reading. If you cannot find their address and/or telephone number, go to the local library and ask to look through their *Thomas Register*. Larger libraries usually have a reference section. Try calling the library by telephone if you only need one or two addresses.

Some manufacturers will put you on their mailing list, especially if you're working in an area of electronics in which their products are used.

If you're a member of a trade or technical society or association, your association will probably print an annual directory that includes a list of manufacturers' names, addresses, and telephone numbers. Sometimes toll-free telephone numbers are also listed along with the names of the key people to contact for information and literature. The ISCET Annual Yearbook is one such directory. Associations such as ISCET also regularly list publications available — sometimes at no cost — to members (FIG. 13-1). Members must provide required postage or copying costs.

QUAN.	NUMBER	TITLE	# OF SOURCE/PAGES	

	MA0587 JE 414	**JOSEPHSON EFFECT — VERIFIED AT 85 DEGREES KELVIN** A new superconducting material, consisting of barium, yttrium, copper and oxygen, has made possible the verification of the Josephson effect at temperatures above that of liquid nitrogen to allow easy interfacing with other widespread LSIs.	Journal of Electronic Engineering	1
	MA0587 JE 415	**EMI COUNTERMEASURE FOR SWITCHING POWER SUPPLIES** Electronic equipment malfunction due to power-line noise can be alleviated with electromagnetic interference countermeasure methods including electronic component selection, filter application, shielding, grounding and layout.	Journal of Electronic Engineering	4
	MA0687 JE 417	**SURFACE-MOUNT DEVICES: CHIP TRIMMER CAPACITORS: THE LATEST TECHNOLOGY AND PRODUCT TRENDS** Trimmer capacitors are adapting to surface-mount technology as they are converted to chip form. Technology, product features and development trends are described.	Journal of Electronic Engineering	3

PLEASE SEND THE REPORTS ITEMIZED ABOVE

Mailing Charges

1-6 pp., 22 cent stamp
7-13 pp., 39 cents postage
Over 13pp., send a second envelope.

Copying Charges

1-4 pp.—Free
5-10 pp.—$1.00
11-20 pp.—$2.00

ENCLOSED ☐ Check ☐ Cash $_____

NAME _____

ADDRESS_____

CITY_____ STATE _____ZIP_____

Send a stamped, self-addressed envelope to:

INTERNATIONAL SOCIETY OF

ISCET

CERTIFIED ELECTRONICS TECHNICIANS

**2708 West Berry Suite 8
Fort Worth, Texas 76109**

Fig. 13-1. A partial example listing some literature regularly available to association members at no cost except for postage or copying charges. (The actual full listing includes 20 items). COURTESY OF NESDA/ISCET.

GOVERNMENT SOURCES

A good low-cost source of books, pamphlets, and other literature on various subjects including electronics is the federal government. Among the items available are military books and technical manuals on various topics such as basic electronics, radar, communications, and so on. The United States Government Printing Office has outlets in larger cities. You can purchase available publications directly at these outlets, or you can purchase them by mail. To get a list of books and other information available, ask for the electronics list by writing to the Superintendent of Documents (address in Appendix F).

BOOKS

Books on various electronics topics can be purchased from local bookstores, or from trade-school or college bookstores. You can also obtain catalogs by calling or writing the publishers of electronics books. Publishers' names can be obtained from the books that you are using in your studies or on the job, some can also be found in Appendix C.

CATALOGS

Catalogs of electronics parts and equipment are available from numerous sources. Some of the catalogs show parts and supplies for the repair of electronic equipment, and for building or assembling experimental circuits. Other catalogs offer discounts on electronic equipment such as VCRs (video cassette recorders), TVs, CBs, hi-fi systems, and so on. If you look through the pages of a magazine such as *Radio Electronics*, you'll see ads from a number of companies offering such catalogs at no cost. In some magazines all you need do is tear out the return card enclosed, check off the numbers corresponding to the companies whose catalogs you would like to have, fill in your address and mail it in. If you're new in electronics, these catalogs can be very educational. They usually show a picture or sketch of each part along with a description. A few of the hundreds of catalogs available are listed in Appendix D.

SERVICE LITERATURE SOURCES

To service and troubleshoot electronic equipment — even your own TV set — it is necessary to have the schematic and service manual on hand (FIG. 13-2). Without it, it's nearly impossible to understand how the circuit should work, where the parts and test points are located, what the values and ratings of the components are, and what can be expected from the equipment.

Where do you obtain service manuals? One place is directly from the manufacturer. If you are an electronics technician or engineer, call them on the phone. They will probably sell you the service manual you need. A quicker way might be to go to a local distributor. Many of them sell service manuals, or packages of a number of service manuals, or a book that covers a series of model numbers that may include the equipment you are working on. To get the right manual, you will need the name of the manufacturer and the model number of the equipment.

One publisher that publishes schematics and service manuals as a major part of its business is the Howard W. Sams Co., Inc. They can provide you with schematics and service manuals for practically any TV set, radio, CB, VCR, or stereo. On request they

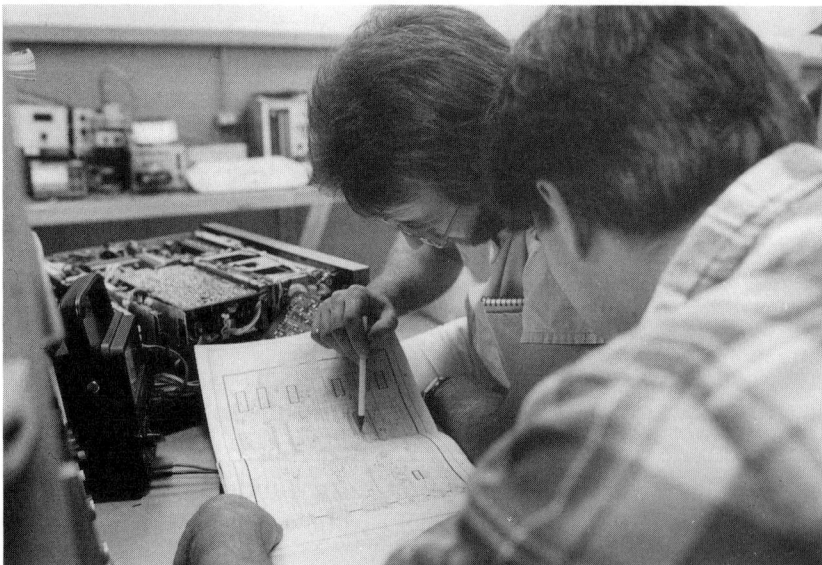

Fig. 13-2. Service literature is essential in analyzing or troubleshooting electronic equipment. COURTESY OF JOHNSON TECHNICAL INSTITUTE.

will send you a catalog listing the model numbers for which they stock service manuals. When you buy a service manual from Sams, the package (or folder) includes the service information you need plus service information on several other equipment models which you probably will not need at that time. However, if you are in the business of repairing electronic equipment, set up a file for these unneeded service manuals. You'll probably need some of them sooner or later.

The Howard S. Sams service literature can also be purchased from certain local distributors of electronic parts. Check by telephone to see if one of the distributors in your area carries the Sams schematics.

A list of some of the more popular manufacturers is in Appendix E.

FREE PUBLICATIONS

You can receive some magazines free, especially if you're already working in some field of electronics, or if you are a member of some associations. For example, if you are a member of the Society of Broadcast Engineers, certain publishers of magazines that cover the field of broadcast engineering will put you on their complimentary subscriptions mailing list. Write to the publisher or to the SBE to obtain the required application. Check with whatever associations you belong to. Ask if any publishers offer free (or reduced-rate) subscriptions.

Many manufacturers and suppliers in electronics work publish pamphlets, brochures, or magazines, called *house organs*, which are available free of charge. Most of them you can get free by writing directly to the firms that publish them. In some cases you might have to be working in the field that the particular publication is intended for. The main purpose of a house organ is to promote the products of the firm publishing it. Even so, many of them include excellent technical articles applying to electronics in general.

The list of electronics equipment manufacturers in Appendix E is a good source for such publications. Write to the ones that interest you. Also, some house organs are listed in Appendix F; however, there are many more.

❖ 14

Advantages of
Association Membership

A LITTLE HAS ALREADY BEEN COVERED ON THE BENEFITS OF MEMbership in an association. Which association you join depends on the type of work you're in. For example, there are associations for cable television, marine radio, broadcast engineering, business and educational radio, consumer electronics, medical electronics, telecommunications, electronic dealers, and space electronics.

Each association provides benefits that will be helpful to its members through information, publications, services, recognition, and promotion of the professional stature of the members to the general public.

If you are already employed, it is possible that your employer will encourage you to join an association and may pay for your membership. Your employer might also pay for you to attend an annual convention of the association, and for seminars dinners, or lunches held in connection with local or regional meetings (FIG.

14-1). Perhaps you'll also be reimbursed for attending upgrading seminars at a regional manufacturer's plant.

Through some associations you can become certified in your particular field of electronics. Appendix G lists major electronics associations. Some of the associations listed (for example, the Electronic Industries Association) do not admit individuals as members. Their membership is made up of manufacturers and other businesses and organizations. However, the EIA is an excellent source of information for technical groups, instructors, publishers, and others. For example, EIA conducts courses on recent developments in electronics in several locations throughout the U.S. during summer months for electronics instructors and other educators. EIA provides films explaining the types of careers available for electronics technicians in electronics. They provide pamphlets on safety and interference, and consumer guides on audio and video products and numerous other topics. EIA compiles data on the range of salaries or pay for electronic technicians. (As of 1985, salaries ranged from $13,000 annually for apprentice and entry positions to $27,000 or more for master

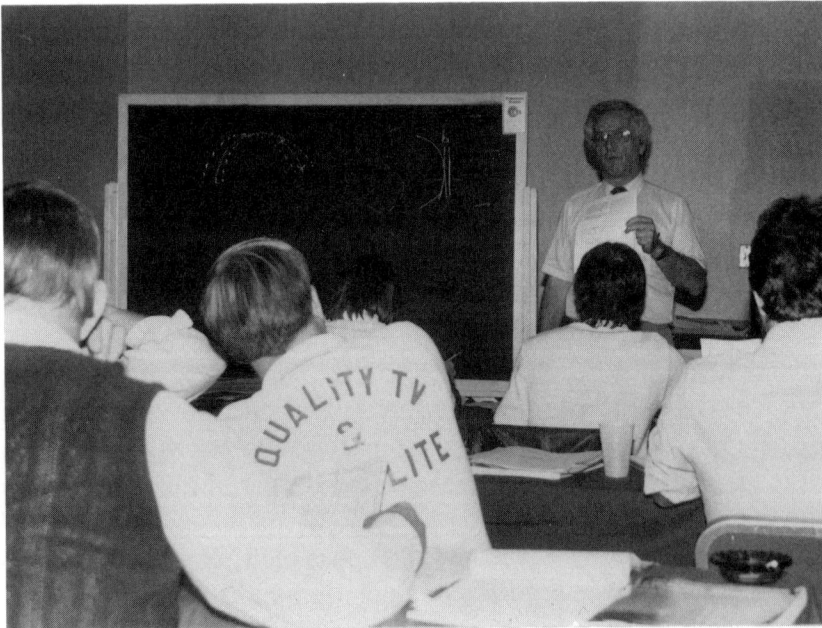

Fig. 14-1. Association seminars are usually informal. Here the instructor is covering the fundamentals of satellite dish polarization. COURTESY OF ELECTRONIC TECHNICIANS ASSOCIATION, INTERNATIONAL.

technicians. Of course, geographic location, the experience and training of the employee, and requirements of the job greatly affect the pay scale for a particular job.)

Most of the associations listed are intended mainly for membership by individuals. If you are going to work in, or if you're already working in the electronic servicing and maintenance or troubleshooting field, you may want to join either ISCET or ETA, or both. You could also join if your work is more closely related to engineering. If you have a good background in mathematics and science, you might also investigate the advantages of certification by NICET, the National Institute for Certification in Engineering Technologies. (Their address is in Appendix G.) NICET is associated with the National Society of Professional Engineers.

Several electronics associations offer various types of certification. For example, ISCET offers certification as a consumer electronics technician. Other ISCET options for certification include audio, communications, computer, industrial, master antenna TV, medical, and radar. ISCET is also permitted by the Federal Communications Commission to issue General radiotelephone licenses by written test.

The Electronic Technicians Association (ETA) also certifies in consumer electronics but with sub-option choices of either radio/TV or audio. Another ETA category with sub-option certification in either two-way radio or avionics is communications. Certification is also available in commercial electronics with sub-option choices of either master antenna systems (MATV) or video tape recorders (VTRs). The remaining ETA options are industrial electronics, computer electronics, and biomedical electronics.

What do you get by joining an association? Her are some of the benefits as provided by ETA, as one example. First you have the strength of numbers. You're a member of a larger group whose purpose is to upgrade the professional technician. The result is increased recognition and should lead to higher wages. In the ETA Technician Association News each month you get circuit or equipment descriptions (for example, satellite transmission and receiving systems) and short quizzes as a self check on your electronics knowledge. Associations have available practice tests or examples of the kinds of questions a person is asked in order to become certified (FIG. 14-2). If you have your own service business, you're eligible to participate in ETA's Success Management Institute business seminars. You get discounts on technical books and trade magazines, and travel/hotel discounts plus savings on car rental. Decals are provided for your shop, office, or auto windows. Low-cost health insurance and accident insurance group policies are available. Also, the association gets technicians together at local and regional meetings, and at annual conventions. It helps technicians find better jobs and helps employers find better technicians.

ELECTRONIC TECHNICIAN CERTIFICATION ADMINISTRATORS

The International Society of Certified Electronics Technicians (ISCET) and the Electronic Technicians Association (ETA) are the two main associations for service technicians of all types includ-

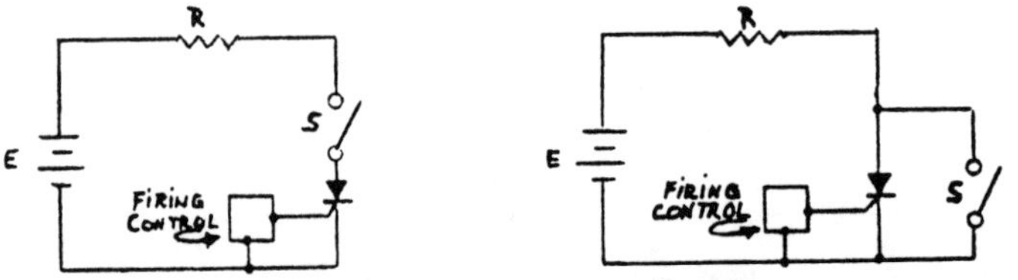

Fig. 14-2. Example of part of a Practice Test showing the kinds of questions in the Industrial Electronic CET exam. COURTESY OF ELECTRONIC TECHNICIANS ASSOCIATION, INTERNATIONAL.

ing consumer, industrial, medical, communications, computer, radar, audio, and others.

A list of certification administrators for ISCET is found in Appendix H, and the ETA list is in Appendix I. A certification administrator is a person who administrates written tests for becoming certified. If you're interested, contact the administrator closest to you. For additional information, write directly to the national headquarters of the association.

BROADCAST TECHNICIAN AND ENGINEER CERTIFICATION

The certification program of the Society of Broadcast Engineers has become recognized by broadcast station owners and managers as a source for qualified technicians and engineers. The SBE offers its certification program to both members and non-members in several grades. Over 2000 of the SBE's 5000 members are currently certified. Certification tests are provided two or more times per year at the society's 90 local chapters. If there is no chapter in your area, the SBE will arrange for you to meet with a proctor.

SBE certification grades are Broadcast Technologist (an entry-level rating), Broadcast Engineer, and Senior Broadcast Engineer. You can qualify as a Broadcast Technologist if you have an FCC first-class radiotelephone license or if you pass the technologist exam.

Five years of experience is the main qualification for the Broadcast Engineer rating. Ten years of experience enables you to apply for the rating of Senior Broadcast engineer. Re-certification is required every five years to help assure that those certified have kept their skills and knowledge up-to-date.

Examination fees are $30 for members and $50 for non-members. The $50 fee includes a one-year membership.

To get more information or a test application, contact a local SBE chapter or the national headquarters. If you are already working in broadcasting or have an interest in broadcast engineers, you can join the SBE even if you're not certified by contacting the address just given. Even if you are not a member of the SBE, however, chapters generally welcome interested non-members at their local meetings. It's a good way to meet others working in the field and to keep up on technical positions and other developments.

A list of the SBE chapters and persons to contact throughout the U.S. plus one in Canada, is provided in Appendix J.

ELECTRONICS ENGINEER AND ENGINEERING TECHNICIAN ASSOCIATION

The Institute of Electrical and Electronics Engineers (IEEE) is an association for engineers and engineering technicians. Local and regional meetings are held throughout the U.S. and there are annual conventions. The IEEE also includes sub-groups, but it does not do any certifying. You can join the sub-groups and receive that sub-group's publications, attend their meetings, and so on. The sub-groups cover various specialties such as industrial electronics, engineering writing and speech, computer science, and instrumentation and control.

A postcard sent to any of the associations listed will bring return information on qualification for and benefits of membership. As a person setting out on or considering a career in electronics, it could be to your benefit to become involved in one or more of the associations. Just to get to know persons who are already successful in electronics as well as others who have conditions and interests similar to yours can be worth the cost of membership.

❖ 15
Changing Jobs or Finding a Job

ONE METHOD OF BETTERING YOURSELF IN ANY FIELD OF WORK, including electronics, is that of *job hopping*, or going from one job to another. Often, if you're already working, you may come across the opportunity to leave your present position and go on to some other job that offers more money, more prestige, greater opportunity for advancement, or greater satisfaction — or a combination of these. Generally, there's nothing wrong with this method so long as you give your present employer sufficient notice that you're leaving and you've fulfilled all obligations. Of course you shouldn't job-hop too often or you may have difficulty explaining to some potential future employer why you cannot hold a job very long or why you get dissatisfied with your work so often.

One thing to consider in leaving a job that you have held for any substantial length of time is whether you're giving up any pension or retirement benefits or vacation time.

EMPLOYMENT AGENCIES

If you have some experience in electronics, such as two or three years or more, and you are interested in going on to a more responsible or better-paying job, you might consider registering with an employment agency. You can find the names of these in the Yellow Pages section of a telephone directory, or among the classified ads of a newspaper. Select an agency that specializes in technical or electronics positions.

Before signing up with an employment agency be sure you understand whether or not there is a fee for their helping you to find a position. In many cases, the fee is paid by the employer. If it is not paid by the employer, you will be expected to pay the agency a certain percentage of your salary for a specified period of time. This could add up to a substantial amount of money, so be sure you understand all the conditions.

TECHNICAL MEETINGS

One way of finding a job or learning of the availability of a better job is by attending local and regional technical meetings and becoming acquainted with other technicians or engineers during the coffee breaks, lunches, or dinners (FIG. 15-1). Many managers, supervisors, head technicians, and engineers attend these meetings and conventions and often are on the lookout for persons to fill positions in their organizations. Some companies send personnel representatives to national and regional technical conventions to interview persons interested in inquiring about working for their company. If you go to such a meeting, look for the employment interviewing area. Sometimes there are bulletin boards listing job openings, and in many cases you can place a notice of your own on the bulletin board listing your qualifications, what kind of position you're looking for and where you can be reached either during the meeting or later. Also, you shouldn't overlook the listings usually found in the monthly publication of your Society or Association, as in the example of FIG. 15-2.

SPREADING THE WORD

If you're looking for a job or a better job, letting people know that you are available can be very helpful. Talk to the sales managers of local electronics distributors. They make contact with local service organizations and industries and often hear in advance of openings that might exist in the near future. Also talk to the people who work the counter at parts distributors and parts supply houses. Technicians and engineers who come in to buy parts and supplies sometimes mention job openings to the sales people at the counter.

Mention at local association meetings that you're available for work or for a different job that offers advancement and greater responsibility. If you feel that you have advanced in knowledge and skill sufficiently to be of significant value to a certain company or organization, you might consider going directly to the owner or manager of that organization and explain just how you think you might be of importance to them in a new position that does not yet exist in their concern. In that way you would be making your own job opening. However, this requires careful planning ahead of time, and some convincing on your part that you could really be a plus to their operation.

Fig. 15-1. Some of those attending this association seminar could represent companies that may be your potential new employer. During the lunch or coffee break, go out of your way to get acquainted. COURTESY OF ELECTRONIC TECHNICIAN ASSOCIATION, INTERNATIONAL.

Another method of spreading the word that you're available is by advertising. Place an ad in your local paper under "Situations Wanted." Have replies sent to a newspaper box number, so that your present employer does not know that you're looking for a change. In fact, if you are already working, mention only to people that you can absolutely trust that you're available for a new job. Otherwise, your present boss may find someone to fill your job before you can find a new one.

JOB LISTINGS FOR TECHNICIANS

The following positions are listed at the ISCET office this month:

Central Intelligence Agency, McLean, VA, seeks electronics technicians with knowledge of RF theory/circuitry, solid state, applications.

General Dynamics, Pomona, CA, seeks test technicians and operators for component-level troubleshooting and testing of PC boards, missile systems and subsystems.

Naval Research Laboratory, Washington, DC, seeks microwave, systems, and general electronics technicians.

Federal Aviation Administration, nationwide, offers openings in various areas for electronics technicians.

Micom Systems, Inc., Simi Valley, CA, seeks experienced senior test technician.

International Education Service, Tokyo, seeks technician to teach engineering English in Japan.

CINDE seeks electrical/electronic engineering professors to teach in Costa Rica.

Better Audio & TV, Elmhurst, IL, seeks experienced bench technician.

Sencore, Sioux Falls, SD, seeks electronics technicians for component-level analyzing and troubleshooting of PC boards.

SFS Corp., Compton, CA, seeks technicians.

NEC Home Electronics, Wood Dale, IL, offers opportunities for experienced technicians.

Mosler Safe Company, Sparks, NV, seeks technician to service electronic alarm equipment.

TESST Electronics School, Hyattsville, MD, seeks electronics/communications instructor.

Honeywell I.S.D., Los Angeles, CA, seeks electronics/computer technicians with hands-on experience.

Show Biz Pizza Time, Inc., Irving, Tx, seeks electronic technician/mechanic with experience to maintain 50-unit game room.

RICH, Franklin Park, IL, seeks technicians for prototyping, troubleshooting video circuits, modifications, and testing.

Curtis Mathes, Jackson, TN, seeks shop manager.

Southern Sound and TV Service, Chattanooga, TN, seeks student training for practical, hands-on experience and troubleshooting.

For more information on these positions contact the

ISCET office: 2708 W. Berry St., Fort Worth, TX 87109;
(817) 921-9101.

Fig. 15-2. Example of job listing taken from ISCET membership publication, UPDATE. COURTESY OF NESDA/ISCET.

PREPARING A RESUMÉ

If you apply for any kind of a position in electronics other than routine production-line assembly or other work not requiring previous experience or education, it's usually a good idea to provide a resumé for your prospective employer. A resumé is a one- or two-page condensation of your personal and technical qualifications. Although an employer cannot legally ask for certain details about you in an employer's application form, you're free to put what you want in your own resumé. For example, things you can include in your resumé, but which an employer cannot legally ask you, is whether you're married or single or divorced, or what your age is.

It's good to have a resumé because you can mail it to an employer, or bring it, before you ever fill out an application. A resumé sometimes gets the attention of a prospective employer whereas your application may receive just a hurried glance.

Since a personnel director or department head usually just skims over an application or resumé for things that might catch his interest, and then goes on to the next application or resumé, a resumé should be only a page or two in length, no matter how much education, training, and experience you've had.

A resumé should include your name, address, and telephone number; your job or position objective; your education; your technical qualifications; your experience; and your levels of responsibility or authority. If you mail a resumé to a prospective employer, include a cover letter mentioning briefly that your resumé is enclosed and describing the position you would like to be considered for. Also, the letter should contain any special qualification that might be of interest to that employer.

Resumés work best if they are individually typed. If this is not practical, have at least one resumé professionally typed and then make good clean photocopies of it.

Resumés vary in style and appearance. A resumé should always be on 8 1/2 × 11 white or ivory paper. The classified ad section of most newspapers include a listing of persons or concerns offering assistance with resumes. If you're attending a school or college, they probably offer a resume assistance service to students who will be graduated within the next four to six months.

For more detailed information on resume preparation plus

```
┌─────────────────────────────────────────────────────────────────┐
│                      Harold A. Wescott                          │
│                  26231 Northern Boulevard                       │
│                   Farmingdale, NY 10312                         │
│                     (516) 555-3232                              │
│                                                                 │
│                                                                 │
│  OBJECTIVE     To obtain an entry-level position in testing and troubleshooting │
│               electronic equipment in a manufacturer's production facility.     │
│                                                                 │
│  EDUCATION     Johnson Technical Institute, Scranton, PA        │
│               Associate in Specialized Technology Degree, Industrial Electronic │
│               Technology                                        │
│               National Technical Schools, Los Angeles, CA       │
│               Diploma in Television Servicing (Home study course) │
│                                                                 │
│  EXPERIENCE    Instrument Installer, Grumman Aircraft Corporation, Bethpage, NY │
│               Installing flight instruments in military aircraft.  Assembling   │
│               plugs and connectors onto cables and using automated checker.     │
│                                                                 │
│  6/85-7/86     Installer Assistant, ACME Electronics, Rahway, NJ │
│               Helper in installing and initial testing of x-ray and diathermy   │
│               equipment in industrial and medical locations throughout New York, │
│               New Jersey, Pennsylvania, and Maryland.           │
│                                                                 │
│  8/84-6/85     Electrician Helper, Triple-A Security, Plainfield, NJ │
│               Prepared materials for electricians, and assisted them in         │
│               installing industrial fire and intruder alarms.   │
│                                                                 │
└─────────────────────────────────────────────────────────────────┘
```

Fig. 15-3. A sample resumé.

RECOGNITION	Received extra day off with pay from Triple-A Security for no lost time on the job. Received bonus from Acme Electronics for suggestion saving company time and money. Passed written test to become Associate Certified Electronic Technician. Certified by Electronic Technicians Association.
SKILLS	Experienced in use of electronic test instruments including digital voltmeter, oscilloscope (triggered, dual sweep), RF signal generator, TV sweep generator, megger, and others.
ADDITIONAL INFORMATION	Own my own personal computer and can program using BASIC language.
PERSONAL INFORMATION	Age: 27. Height: 6'1". Weight: 192 lbs. Health: excellent, married, 2 children ages 4 and 2.
EMPLOYMENT AND PERSONAL REFERENCES	Available on request.

```
                                      January 23, 1988


                                      26231 Northern Boulevard

                                      Farmingdale, NY 10312



Mr. Gerald Siebert

Employment Manager

Third-Party Electronic Maintenance

135 Henry Street

Binghamton, NY 19131


Dear Mr. Siebert:

In response to your ad in the Binghamton Press for an electronics customer

service engineer, please refer to my resume enclosed.


I feel that I could carry out this type of work to your complete satisfaction.

I've been told by previous employers that I've been especially valuable to them

when dealing with customers directly.  I've also gotten a number of commendations

for my knowledge and troubleshooting of my employers' products.


I'm hopeful that I will hear from you shortly as to when I can visit your company

to learn more of the available position and provide further information about

myself.


Thank you.

                                      Sincerely,

                                      Harold A. Wescott
```

Fig. 15-4. A sample cover letter.

placement help in obtaining a new position, write to Scientific Placement and ask for their Resume Wookbook and Career Planner. This organization offers assistance at no cost to engineers and engineering technicians. Their address is in Appendix G.

An example of a resume is shown in FIG. 15-3.

COVER LETTER

If you are sending a resumé to a prospective employer, it should not go all by itself. There should be a cover letter enclosed, stating why you are sending the resumé, what kind of work you are interested in, and pointing out additional qualifications that are not covered in the resumé or special reasons why your services would be of value to that employer. The cover letter could also emphasize in greater detail things that are already covered in the resume. However, the cover letter should not be lengthy. It should not take more than a minute or so to read over.

FIGURE 15-4 shows one example of a cover letter that could accompany the resumé in FIG. 15-3.

❖ Appendix A
Educational Resources

TRAINING AIDS

American Radio Relay League
Newington, CT 06111

Chaney Electronics, Inc.
P.O. Box 4116
Scottsdale, AZ 85261

Edmund Scientific
101 East Gloucester Pike
Barrington, NJ 08007

Electronic Industries Association
2001 Eye St. NW
Washington, DC 20006

Elenco Electronics
616 S. Wheeling Ave.
Wheeling, IL 60090

EMCO Electronics
104 S. Central Ave.
Valley Stream, NY 11580

Hands-On Electronics
Subscription Department
P.O. Box 338
Mt. Morris, IL 61054

Kelvin Electronics Inc.
P.O. Box 8, 1900 New Highway
Farmingdale, NY 11735

McGraw-Hill Continuing Education
3939 Wisconsin Ave., NW
Washington, DC 20016

RISSCO
P.O. Box 131
Dunmore, PA 18512

Tesla Coil Builders Association
R.D. #3, Box 181, Army Lane
Glens Falls, NY 12801

INFORMATION ON APPRENTICESHIPS

International Brotherhood of Electrical Workers
Department of Skills Improvement
1125 15th, N.W.
Washington, DC 20005

National Joint Apprenticeship and Training
Committee for the Electrical Industry
9300 George Palmer Highway
Lanham, MD 20801

National Electrical Contractors Association
7315 Wisconsin Ave.
Washington, DC 20014

Edison Electric Institute
1140 Connecticut Ave. NW
Washington, DC 20036

Electrical Apparatus Service Association, Inc.
1311 Bauer Blvd.
St. Louis, MO 63132

U.S. Department of Labor
Bureau of Apprenticeship and Training
601 D Street, NW
Washington, DC 20213

CORRESPONDENCE SCHOOLS

American Council on Education
One DuPont Circle NW
Washington, DC 20036

Cleveland Institute of Electronics
1776 East 17th St.
Cleveland, OH 44144

Grantham College of Engineering
12100 Grandview Rd.
Grandview, MO 64030

Granton Institute of Technology
263 Adelaide St., West
Toronto, Canada M5H 1Y3

Heathkit/Zenith Educational Systems
Hilltop Rd.
St. Joseph, MI 49085

Hemphill Schools
510 S. Alvarado St.
Los Angeles, CA 90057

ICS—International Correspondence Schools
Scranton, PA 18515

ICS Center for Degree Studies
Scranton, PA 18515

McGraw—Hill Continuing Education Center
3939 Wisconsin Ave., NW
Washington, DC 20016

NAB
1771 N St.
Washington, DC 20036

National Association of Trade and
Technical Schools (NATTS)
2251 Wisconson Ave. NW
Washington, DC 20007

National Home Study Council
1601 18th St NW
Washington, DC 20009

National Technical Schools
4000 Figueroa St.
Los Angeles, CA 90037

National Textbook Company
VCM Career Horizons
4255 West Touby Ave.
Lincolnwood, IL 60646

NRI Schools
3939 Wisconsin Ave., NW
Washington, DC 20016

❖ Appendix B
Electronics Publications

APCO Bulletin
105½ Canal St.
New Smyrna, FA 32070

Appliance Service News
PO Box 789
Lombard, IL 60148

Audio Video
Dempa Publications
400 Madison Ave.
New York, NY 10017

Audio Visual Communications
Media Horizons, Inc.
50 West 23rd St.
New York, NY 10000

Autosound and Communications
135 West 50th St.
New York, NY 10020

Broadcast Engineering
Intertec Publishing Corporation
9221 Quivera Road
Overland Park, KS 12901

Business Radio
NABER
1330 New Hampshire Ave. NW
Washington, DC 19164

BYTE
Suite 21, Strand Building
174 Concord St.
Peterborough, NH 03459

CEE (Electrical Construction)
Sutton Publishing Company
707 Westchester Ave.
White Plains, NY 10604

Communications
Cardiff Publishing Company
6530 S. Yosemite St.
Englewood, CO 80111

Computer Decisions
10 Mulholland Dr.
Hasbrouck Heights, NJ 07604

Computer Design
119 Russell St.
P.O. Box 593
Littleton, MA 01460

Computer/Electronic Service News
Suite 21, Strand Building
174 Concord St.
Peterborough, NH 03458

CB Radio Times
2278 Industrial Blvd. #101
Norman, OK 73069

Computer World
Suite 21, Strand Building
174 Concord St.
Peterborough, NH 03458

Dealerscope
115 Second Ave.
Waltham, MA 02154

Digital Audio
WGE Center
PO Box 976
Farmingdale, NY 11737

The Electron
Cleveland Institute of Electronics
4781 East 355th St.
Willoughby, OH 44084

Electronic Engineering Times
CMP Publications
600 Community Dr.
Manhassett, NY 11630

Electronic Design
10 Mulholland Dr.
Hasbrouck Heights, NJ 07604

Electronic Technicians Association (ETA)
News
825 East Franklin St.
Greencastle, IN 46135

Electronic Education
1311 Executive Center Dr.
Suite 220
Talahassee, FL 32301

Electronic Packaging and Production
Cahners Publishing
275 Washington St.
Newton, MA 02158

Electronic Servicing and Technology
PO Box 12901
Overland Park, KS 66212

Electronic Service News
174 Concord St.
Peterborough, NH 03458

Electronic Technician Dealer
757 Third Ave.
New York, NY 10017

Electronics
McGraw-Hill, Inc.
1221 Avenue of the Americas
New York, NY 10017

Electronics Test
1050 Commonwealth Ave.
Boston, MA 02215

IEEE Spectrum
Institute of Electrical & Electronics
Engineers
345 E. 47th St.
New York, NY 10017

Ham Radio
Greenville, NH 03048

Industrial Research and Development
Technical Publishing
PO Box 1030
1301 S. Grove Ave.
Barrington, IL 60010

Instrumentation and Control News
PO Box 2005
Radnor, NJ 19089

Medical Electronics
2992 West Liberty Ave.
Pittsburgh, PA 15216

LASER FOCUS
PennWell Publishing Company
119 Russell St.
Littleton, MA 01460

Microwave System News
PO Box 50249
Palo Alto, CA 94303

Modern Electronics
6855 Santa Monica Blvd.
Suite 200
Los Angeles, CA 90038

Ed Noll
PO Box 75
Chalfont, PA 18914

Personal Communications
4005 Williamsburg Court
Fairfax, VA 22032

Personal Electronics
6255 Barfield Road
Atlanta, GA 30328

PMRSS
N. Assoc. of Business and Education Radio
1300 New Hampshire Ave. N.W.
Washington, DC 20036

Pro Sound News
220 Westbury Ave.
Carle Place, NY 11514

Popular Computing
70 Main St.
Peterborough, NH 03458

Professional Electronics
ISCET
2708 W. Berry St.
Ft. Worth, TX 76109

QST
ARRL
Newington CT 06111

Radio Communications Reporter
1701 K St. NW
Washington, DC 20006

Radio Electronics
500-B BiCounty Blvd.
Farmingdale, NY 11735

Radio World
PO Box 1214
Falls Church, VA 22041

RCA Communicator
600 N. Sherman Dr.
Indianapolis, IN 47201

Robotics Age
174 Concord St.
Peterborough, NH 03458

SBE Signal
Society of Broadcast Engineers
7002 Graham Road, Suite 118
Indianapolis, IN 46220

Satellite Retailer
501 N. Washington St.
Shelby, NC 28150

SATVISION
300 N. Washington St., Suite 310
Alexandria, VA 22314

SMPTE Journal
Society of Motion Picture & TV Engineers
862 Scarsdale Ave.
Scarsdale, NY 10583

Sight & Sound
Drobaugh Publications
51 E. 42nd St.
New York, NY 10017

STV (Satellite Television)
501 N. Washington St.
Shelby, NC 28151

S&VC (Sound and Video Contractor)
Intertec Publishing Corporation
9221 Quivera Road
Overland Park, KS 66212

Sound and Communications
156 E. 37th St.
New York, NY 10016

Telecommunication Technology
119 Russell St.
Littleton, MA 01460

Television/Broadcast Communications
Globecom Publicating Limited
4121 West 83rd St.
Prairie Village, KS 66208

Television Digest
1836 Jefferson Place
Washington, DC 20036

Television/Radio Age
1270 Avenue of the Americas
New York, NY 10020

Test and Measurement World
2215 Brighton Ave.
Boston, MA 02134

Telephony
53 West Jackson Blvd.
Chicago, IL 60604

Two-Way Radio Dealer
PO Box 4300 TA
Denver, CO 80217

United Technical Publications
645 Stewart Ave.
Garden City, NY 11530

Video Systems
PO Box 12901
Overland Park, KS 66212

❖ Appendix C
Publishers of
Books on Electronics

Academic Press, Inc.
111 Fifth Ave.
New York, NY 10003

American Radio Relay League
Newington, CT 06111

American Technical Publishers, Inc.
12235 S. Laramie Ave.
Alsip, IL 60658

Davis Publications, Inc.
229 Park Ave. South
New York, NY 10003

Delmar Publishers
50 Wolf Road
Albany, NY 12205

Electronic Industries Association
2001 Eye St.
Washington, DC 20006

Gernsback Publications
500-B Bi County Boulevard
Farmingdale, NY 11735

Gregg/McGraw Hill
1221 Avenue of the Americas
New York, NY 10020

Hayden Publishing Company
PO 13803
Philadelphia, PA 19101

Howard W. Sams and Company, Inc.
4300 West 62nd St.
Indianapolis, IN 46268

IEEE Press
345 East 47th St.
New York, NY 10017

Kiver Publications, Inc.
222 W. Adams St.
Chicago, IL 60060

MacLean—Hunter Publishing Corp.
481 University Ave.
Toronto, Ontario, Canada

Macmillan Company
60 Fifth Ave.
New York, NY 10017

McGraw-Hill Book Company
1221 Avenue of the Americas
New York, NY 10020

National Textbook Company
4255 W. Touhy Ave.
Lincolnwood, IL 60646

Noll Publications
PO Box 75
Chalfont, PA 18914

Parker Publishing Company, Inc.
West Nyack, NY 10994

Prentice-Hall, Inc.
Englewood Cliffs, NJ 07632

Radio Shack
1300 One Tandy Center
Ft. Worth, TX 76109

SBE Publications
PO Box 20450
Indianapolis, IN 46250

Sagamore Publishing Company, Inc.
1120 Old Country Road
Plainview, NY 11803

Science Research Associates, Inc.
155 N. Wacker Drive
Chicago, IL 60606

Society of Manufacturing Engineers
Video Communications Dept.
One SME Drive
Dearborn, MI 48121

Technical Publishing Division
National Education Corp.
Oak and Pawnee
Scranton, PA 18515

TAB Books
Blue Ridge Summit, PA 17214

Texas Instruments
PO Box 225621
Dallas, TX 75242

Wadsworth Publishing Company
Ten Davis Drive
Belmont, CA 94002

Wiley Publications
Box 092
Somerset, NJ 08873

John Wiley & Sons, Inc.
One Wiley Drive
Somerset, NJ 08873

Ziff-Davis Publishing Corp.
One Park Ave.
New York, NY 10016

❖ Appendix D
Information and Catalogs

Allied Electronics
401 East 8th St.
Fort Worth, TX 76102

AMCO McLean Corporation
786 McLean Ave.
Yonkers, NY 10704

Analog Devices
Route 1, Industrial Park
PO Box 280
Norwood, MA 02062

Automated Production Equipment
142 Peconic Ave.
Medord, NY 11763

Barnard Electronics
PO Box 15214
Red Bank, TN 37415

Black Box Corporation
PO Box 12800
Pittsburgh, PA 15241

Contact East
335 Willow St. South
PO Box 786
North Andover Business Park, MA 01845

Consolidated Electronics
705 Watervliet Ave.
Dayton, OH 45420

Dempa Publications, Inc.
400 Madison Ave.
New York, NY 10017

Diehl
661 Canyon Drive F
Amarillo, TX 79110

EMCO Electronics
104 S. Central Ave.
Valley Stream, NY 11580

Fordham
260 Motor Parkway
Hauppauge, NY 11788

Graymark
Box 5020
Santa Ana, CA 02704

Heath
Benton Harbor, MI 49022

Hewlett—Packard Company
3000 Hanover St.
Palo Alto, CA 94304

Howard W. Sams and Co., Inc.
4300 West 62nd St.
Indianapolis, IN 46268

Leader Instruments
380 Oser Ave.
Hauppauge, NY 11788

National Semiconductor
Applications Department
2900 Semiconductor Drive
Santa Clara, CA 05051

Newark Electronics
500 N. Pulaski Road
Chicago, IL 60624

Oliger Electronics
11601 Wedbey Drive
Cumberland IN 46229

Polyline Corporation
1233 Rand Road
Des Plaines, IL 60616

Projector/Recorder Belts
2000 Clay St.
Whitewater, WI 53190

Radio Shack
(Check telephone directory for local
retail store), or
500 One Tandy Center
Ft. Worth, TX 76102

G Russell Electronics
Route 1 Box 539
Center Hall, PA 16828

Sencore
3200 Sencore Drive
Sioux Falls, SD 57107

Signetics
Applications Division
811 East Arques Ave.
Sunnyvale, CA 94088

Tentel Corp
1506 Dell Ave.
Campbell, CA 95008

T/S Connection
Attn: J. Roberts
3832 Watterson Ave.
Cincinnati, OH 45227

Unity Electronics
PO Box 213
Elizabeth, NJ 07206

Universal Semiconductor, Inc.
1925 Zanker Road
San Jose, CA 95112

VIZ Test Equipment
335 E. Price St.
Philadelphia, PA 19144

Zebra Systems, Inc.
78-06 Jamaica Ave.
Woodhaven, NY 11421

❖ Appendix E
Consumer Electronics Manufacturers

AKAI American, Ltd.
800 West Artesia Blvd.
Compton, CA 90220
(213) 537-3880

Apple Computer
20525 Mariani Ave.
Cupertino, CA 05130
(408) 996-1010

Atari Corp.
PO Box 3427
Sunnyvale, CA 94088
(408) 745-2000

B&K Precision Test Instruments
Dynascan Corp.
6400 West Cortland
Chicago, IL 60635
(312) 889-9087

Capehart Parts Division
PO Box 127
Voluntown, CT 06384
(203) 376-9932

Channel Master
Industry Park Drive
Smithfield, NC 27577
(919) 934-9711

Commodore International, Ltd.
1200 Wilson Drive
West Chester, PA 19380
(215) 431-9100

Curtis Mathes Corp.
PO Box 223607
Dallas, TX 75222
(800) 527-7646

Electro-Voice
606 Cecil St.
Buchanan, MI 49107
(616) 695-6831

Emerson Radio Corp.
1 Emerson Lane
Seacaucus, NJ 07094
(201) 865-4343

Fisher Corp.
21314 Lassen St.
Chatsworth, CA 91311
(213) 998-7322

John Fluke Mfg. Co.
PO Box C9090
Everett, WA 98206
(206) 342-6100

GC Electronics
400 S. Wyman St.
Rockford, IL 61101
(815) 968-9661

General Electric Company
Video and Audio Operations
Portsmouth, VA 23705
(804) 483-5000

General Instrument Corp.
RF Systems Division
4229 S. Fremont
Tucson, AZ 85714
(900) 847-0721

General Motors Corp.
Delco Electronics Division
700 East Firmin St.
Kokomo, IN 46901
(317) 459-1194

Gold Star Electronics
1050 Wall St. W
Lyndhurst, NJ 07071
(201) 460-8870

Heath
PO Box 1288
Benton Harbor, MI 49022
(616) 982-3200

Hewlett Packard
3000 Hanover St.
Palo Alto, CA 94304
(415) 857-1501

Hitachi
401 West Artesia Blvd.
Compton, CA 90220
(213) 537-8383

IBM Corp.
PO Box 2989, Bldg. 239
Del Ray Beach, FL 33444
(305) 241-7747

Ikegami Electronics (USA)
37 Brook Ave.
Maywood, NJ 07607
(201) 368-9171

JVC, Division, U.S. JVC Corp.
41 Slater Drive
Elmwood Park, NJ 07407

E.F. Johnson Company
206 2nd Ave. SW
Waseca, MN 56093
(507) 835-6371

Kenwood Electronics
1315 E. Watsoncenter Road
Carson, CA 90745
(213) 518-1700

Kikusui International Corp.
17819 S. Figueroa St.
Gardena, CA 90248

Leader Instruments Corp.
380 Oser Ave.
Huppauge, NY 11788
(516) 231-6900

Marantz Company, Inc.
20525 Nordhoff St.
Chatsworth, CA 91311
(818) 998-9333

Matsushita Engineering and Service Co.
50 Meadowland Ave.
Secaucus, NJ 07094
(201) 348-7000

Midland International
1690 N. Topping
Kansas City, MO 64120
(816) 241-8500

Mitsubishi Electric Sales
3019 E. Maria
Compton, CA 90221
(213) 537-7132

NAP Consumer Electronics Corp.
PO Box 555
Old Andrew Johnson Hwy.
Jefferson City, TN 37760
(615) 475-3801

NEC Home Electronics, U.S.A.
1401 Estes Ave.
Elk Grove Village, IL 60007
(312) 228-5900

New-Tone Electronics, Inc.
44 Farrand St.
Bloomfield, NJ 07003
(201) 748-5089

North Star Computers, Inc.
1440 Catalina St.
San Leandro, CA 94577
(415) 357-8500

Okidata
532 Fellowship Road
Mount Laurel, NJ 08054
(609) 235-2600

Panasonic
50 Meadowland Ave.
Secaucus, NJ 07094
(201) 348-7000

PTS Corporation
PO Box 272
Bloomington, IN 47402
(812) 824-9931

Philco
PO Box 555
Old Andrew Johnson Hwy.
Jefferson City, TN 37760

Philips ECG, Inc.
100 First Ave.
Waltham, MA 02254

Philips Electronic Instruments
85 McKee Ave.
Mahwah, NJ 07430
(201) 529-3800

Pioneer Electronics, Inc.
5000 Airport Plaza Drive
Long Beach, CA 90801
(213) 420-5700

Quasar
50 Meadowland Parkway
Secaucus, NJ 07094
(201) 348-7000

RCA Consumer Electronics
600 N. Sherman Drive
Indianapolis, IN 46201
(317) 265-5000

RCA Distributor/Special Products
2000 Clements Bridge Road
Deptford, NJ 08096
(609) 541-3636

RCA Data Communications
New Holland Ave.
Lancaster, PA 17604

Radio Shack
1300 One Tandy Center
Ft. Worth, TX 76102
(817) 390-3011

Sansui Electronics Corp.
17150 S. Margay Ave.
Carson, CA 90746
(213) 604-7300

Sanyo Electric, Inc.
1200 W. Artesia Blvd.
Compton, CA 90220
(213) 537-5830

H.H. Scott
20 Commerce Way
Woburn, MA 01888
(617) 933-8800

Sears
Sears Tower
Chicago, IL 60684
(312) 875-5222

Sencore, Inc.
3200 Sencore Drive
Sioux Falls, SD 57107
(605) 339-0100

Sharp Electronics Corp.
2 Sharp Plaza
Paramus, NJ 07652
(201) 265-5600
17107 Kingsview Avenue
Carson, CA 90746
(213) 515-6866

Simpson Electric
853 Dundee Ave.
Elgin, IL 60120
(312) 697-2260

SONY Corp. of America
Sony Drive
Park Ridge, NJ 07656
(201) 930-1000

Sound Design
34 Exchange Place
Jersey City, NJ 07302
(201) 434-2050

Sperry Instruments
245 Marcus Blvd.
Hauppauge, NY 11788
(516) 231-7050

Sprague Electric Co.
551 Marshall St.
North Adams, MA 01247
(413) 664-4481

TEAC Corp. of America
7733 Telegraph Road
Montebello, CA 90640

Tech Spray, Inc.
PO Box 949
Amarillo, TX 79105
(806) 372-8523

Tektronix, Inc.
PO Box 500
Beaverton, OR 97077

Texas Instruments
PO Box 225621
Dallas, TX 75243
(214) 995-2011

Thordarson Meissner, Inc.
Mt. Carmel, IL 62863

Toshiba America, Inc.
82 Totowa Road
Wayne, NJ 07470
(201) 628-8000

Triplett Instruments
One Triplett Drive
Bluffton, OH 45817
(419) 358-5015

VIZ Test Equipment
325 E. Price St.
Philadelphia, PA 19144
(215) 844-2626

Wang Laboratories
One Industrial Ave.
Lowell, MA 01851
(617) 459-5000

Wells Gardner Electronics Corp.
2701 N. Kildare
Chicago, IL 60639
(312) 252-8220

Western Auto Supply Company
2107 Grand Ave.
Kansas City, MO 64108
(816) 346-4001

Weston Instruments
614 Freylinghuysen Ave.
Newark, NJ 07114
(201) 242-2100

Workman Electronic Products
75 Packinghouse Road
Sarasota, FL 30578
(813) 371-4242

Yamaha Electronics Corp.
6666 Orangethorpe
Buena Park, CA 90620
(714) 522-9105

Zenith Data Systems
1000 Milwaukee Ave.
Glenview, IL 60025
(312) 391-8860

Zenith Videotech Corp.
1100 Seymour Ave.
Franklin Park, IL 60131
(312) 671-7550

❖ Appendix F
Free
Electronics Publications

B & K Precision Test Instruments
Dynascan Corporation
6460 West Cortland
Chicago, IL 60635

DIEHL
6661 Canyon Drive F
Amarillo, TX 79110

Digital Equipment Corp.
146 Main St.
Maynard, MA 01754

Electro Review
Technical Information Distribution
Service
40 West Ridgewood Ave.
Ridgewood, NJ 07451

E.I.L. Instruments, Inc.
10 Loveton Circle
Sparks, MD 21152

IES News
Illuminating Engineering Society of
North America
345 E. 47th St.
New York, NY 10017

Fairchild Digital Logic Division
333 Western Ave.
South Portland, ME 04106

John Fluke Mfg. Co., Inc.
PO Box C9090
Everett, WA 96206

General Electric
Electronics Park
Syracuse, NY 19132

Heath
PO Box 1288
Benton Harbor, MI 49022

Hewlett-Packard Journal
Hewlett-Packard Company
3000 Hanover Street
Palo Alto, CA 94304

E.F. Johnson Company
206 2nd Ave. SW
Waseca, MN 56093

Midland International
1690 N. Topping
Kansas City, MO 64120

Philips ECG, Inc.
Distributor & Special Marketing Division
100 First Ave.
Waltham, MA 02254

RCA Corporation
Consumer Electronics Technical
Publications
600 N. Sherman Drive
Indianapolis, IN 46201

RCA Data Communications Products
New Holland Ave.
Lancaster, PA 17604

RCA Distributor and Special Products
Division
2000 Clements Bridge Road
Deptford, NJ 08096

Howard W. Sams & Co., Inc.
4300 West 62nd St.
Indianapolis, IN 46268

Semiconductor Data Update
Motorola
PO Box 20912
Phoenix, AZ 85036

Sencore News
Sencore
3200 Sencore Drive
Sioux Falls, SD 57107

Sprague Electric Company Division
551 Marshall St.
North Adams, MA 01247

Thordarson Meissner, Inc.
Electronic Center
Mount Carmel, IL 62863

Tech Spray, Inc.
PO Box 949
Amarillo, TX 79105

Tektronix, Inc.
PO Box 500
Beaverton, OR 97077

Test and Measurement World
199 Wells Ave.
Newton, MA 02159

TESLA CO.
Suite 6, 490 S. Rosemeade Blvd.
Pasadena, CA 91107

Texas Instruments Product Bulletin
Texas Instruments Incorporated
PO Box 809066
Dallas, TX 75380

Triplet
One Triplett Drive
Bluffton, OH 45817

Universal Semiconductor, Inc.
1925 Zanker Road
San Jose, CA 95112

U.S. Government Printing Office
Superintendent of Documents
Pueblo, CO

Workman Electronic Products, Inc.
75 Packinghouse Road
PO Box 3628
Sarasota, FL 33578

Zenith Data Systems
1000 Milwaukee Ave.
Glenview, IL 60025

ZIATECH Corporation
3433 Roberto Court
San Luis Obispo, CA 93401

Zenith Videotech Corp.
11000 Seymour Ave.
Franklin Park, IL 60131

❖ Appendix G
Associations and Professional Groups

American Electronics Assoc. (AEA)
2670 Hanover St.
Palo Alto, CA 94304

American Radio Relay League (ARRL)
Newington, CT 06111

Armed Forces Communications and
Electronics Association (AFCEA)
5641 Burke Centre Parkway
Burke, VA 22015

Association of Communications
Technicians (NABER-ACT)
PO Box 19164
Washington, DC 20036

Associated Public-Safety
Communications Officers (APCO)
PO Box 669
New Smyrna Beach, FL 32070

Direct Broadcast Satellite Association
(DBSA)
1800 M St. NW, Suite 400
Washington, DC 20036

Electronic Industries Association
2001 Eye St. NW
Washington, DC 20006

Energy Telecommunications and
Electrical Association
PO Box 795038
Dallas, TX 75379

Electronic Technicians Association
International (ETA, ETA-I)
825 East Franklin St.
Greencastle, IN 46135

Engineer's Council for Professional
Development
345 East 47th St.
New York, NY 10017

Institute of Electrical and Electronics
Engineers (IEEE)
345 East 47th St.
New York, NY 10017

International Society of Certified
Electronics Technicians (ISCET)
2708 West Berry St.
Fort Worth, TX 76109

Japan Electronics Bureau
One Penn Plaza
250 West 34th St., Suite 1533
New York, NY 10119

National Association of Business and
Educational Radio (NABER)
1330 New Hampshire Ave. NW
Washington, DC 20036

National Association of Radio and
Telecommunications Engineers, Inc.
(NARTE)
PO Box 15029
Salem, OR 07309

National Cable Television Association
(NCTA)
1724 Massachusetts Ave. NW
Washington, DC 20036

National Electronic Service Dealers
Association (NESDA)
2708 West Berry
Fort Worth, TX 76109

National Institute for Certification in
Engineering Technologies (NICET)
1420 King St.
Alexandria, VA 23314

National Marine Electronics Association
(NMEA)
PO Box 57
Oronoco, MN 55960

Radio Club of America (RCA)
PO Box 2112
Grand Central Station
New York, NY 10017

Society of Broadcast Engineers (SBE)
7002 Graham Road
Indianapolis, IN 46220

Society of Cable Television Engineers
(SCTE)
PO Box 2389
West Chester, PA 19380

Society of Motion Picture and Television
Engineers (SMPTE)
595 West Hartsdale Ave.
White Plains, NY 10607

SPACE — The Satellite TV Industry
Association
300 North Washington St.
Alexandria, VA 22314

Society for Technical Communication
(STC)
815 15th St. NW
Washington, DC 20005

Tesla Coil Builders Association
R.D. 3, Box 181 Amy Lane
Glens Falls, NY 12801

Telephone Retailers Association
3619 Piedmont Road
Atlanta, GA 30305

United States Telephone Association
(USTA)
900 19th St. NW, Suite 800
Washington, DC 20006

❖ Appendix H
ISCET Certification Administrators

ALABAMA
Hanceville	Leeland Blackmon	(205) 352-6403 ext. 60
Mobile	Cecil F. Edwards	(205) 666-4837
Muscle Shoals	Ricky G. Reaves	(205) 381-2813
Sumiton	Richard Stacks	(205) 648-3271 ext. 39

ALASKA
Fairbanks	Larry Hart	(907) 479-7238

ARIZONA
Cave Creek	Edward J. DeFonzo	(602) 488-4452
Chandler	James Smith	(602) 821-9376
Phoenix	Frank Grabiec	(602) 279-4176
	Russell Scarpelli	(602) 979-5403
	Richard Tisor	(602) 264-1726
Tucson	Robert Madigan	(602) 623-1991

ARKANSAS

Camden	Darrell Cole	(501) 574-0741 ext. 44
Little Rock	Ralph Canterbury, Jr.	(501) 664-0461

CALIFORNIA

Anaheim	Edward N. Wollacott	(714) 634-4565
Barstow	Paul Kovatch	(619) 252-2411
Beverly Hills	Keith Wolf	(213) 271-6446
Carmichael	Donald Surette	(916) 489-1873
Chino	Jerry Bunnell	(714) 627-5708
Concord	Anthony Haag	(415) 686-5636
Fresno	Armen Karagosian	(209) 277-1801
	Michael Maxwell	(209) 449-3512
Glendale	Jim White	(818) 246-2574
Hawthorne	CSEA	(213) 679-9186
Kenfield	James Staley	(415) 485-9530
Long Beach	Joseph Provenza	(213) 421-5996
Lomita	Ellis R. Driver	(213) 326-1414
Los Angeles	Thomas Burkhead	(213) 527-5830 ext. 367
	Bill Lawler	(213) 257-7997
Midway	Fred Potter	(714) 891-9211
North Hollywood	Howard Bardach	(818) 984-1768
Oroville	Ed Strother	(916) 895-2482
Oxnard	Edward Aldridge	(805) 985-6557
	Jim Swigart	(805) 487-2756
Pasadena	Edward Belsky	(213) 551-1155
Rancho Cordova	Jon R. Sturz	(916) 635-6000
Rialto	Dorr Stuart	(714) 686-8543
San Bernadino	Carl T. Leever	(714) 862-3028
San Diego	Donald Latimer	(619) 278-3033
San Gabriel	Silvino Alonso	(818) 286-5500
San Jose	Clifford Hansen	(408) 379-7670
San Luis Obispo	Frederick Smith	(805) 543-2700
Santa Barbara	Peter Gledhill	(805) 687-0114
Sherman Oaks	Charles Alvarez	(818) 788-7141
Smartville	Don Winchel	(916) 639-2477
Van Nuys	Robert Johnson	(818) 901-7311
	Joel Moskowitz	(818) 980-5212
	Hal Robbins	(818) 782-9462
	Ray S. Spikes	(818) 785-6567

Ventura	Earl Loughboro	(805) 642-0241
Westlake Village	Melvyn Weiss	(818) 706-2228

COLORADO

Aurora	John F. Stackhouse	(303) 695-1913
Cortez	Anthony Valdez	(303) 565-7415
Denver	Bobby G. Allen	(303) 426-1808
	James Haigler	(303) 426-1808
	Bud Izen	(303) 371-3060
Grand Junction	Charles Fetters	(303) 248-1365
	Ronald Ismay	(303) 248-1365
Loveland	Art Sorensen	(303) 669-0883
Westminster	Mike Minamoto	(303) 466-8811
Widefield	George Shaiffen	(303) 392-1000

CONNECTICUT

Simsbury	Richard Steg	(203) 651-3942

DELAWARE

Cheswold	Albert Moses	(302) 571-5447
Dover	Ernest Shannon	(302) 734-7950

FLORIDA

Bradenton	Wes Bazell	(813) 756-2245
Clearwater	James Hite	(813) 531-3531
Ft. Lauderdale	Hamilton Boyd	(305) 764-4935
	Denis Falcomer	(305) 476-5410
	Robert M. Foreman	(305) 491-7171
	Edward Guary, Sr.	(305) 463-2998
	Edward Guary, Jr.	(305) 763-2964
Ft. Walton Beach	Wesley Johnstone	(904) 244-5936
Gainesville	John Bishop, Jr.	(904) 376-9955
Jacksonville	Karl Hunter	(904) 744-8480
Largo	George Gluze	(813) 581-3040
Margate	John Scheller	(305) 972-7359
Miami Springs	Hallock Neshe	(305) 888-7369
Orlando	Perry Parks	(305) 699-8811
Pensacola	Harold S. Martin	(305) 476-5410 ext.'s 2310-2317
Riverview	Thomas Elmore	(813) 677-1824
St. Petersburg	Ronald DeBoy	(813) 327-3671

Satellite Beach	Walter Lane	(305) 777-1728
Sebastian	Milton Farber	(305) 589-3564
Tampa	William Brooks	(813) 238-0455
	Hilton J. Day	(813) 879-8333
	James Etienne	(813) 884-0024
	James Russell	(813) 857-3717
Titusville	Denis Abell	(305) 269-4208
West Palm Beach	Noel Shevack	(305) 842-8324
Winter Haven	Cal Reddick	(813) 299-6514

GEORGIA
Americus	W.R. Rooks	(912) 928-0283
		(912) 924-4704
Clarkston	Del Agerton	(404) 296-9544
Clarksville	Roy Chastain	(404) 754-2131 ext. 334
Macon	Ebin Shepard	(912) 788-1187
Norcross	James Peek	(404) 925-6858
Powder Springs	D.R. Bosley	(404) 943-8948
Scotdale	Daniel B. Mundy	(404) 297-7378
Statesboro	Dr. Thomas Singletary	(912) 681-5112
Swainsboro	Joe Reese	(912) 237-7010

HAWAII
Kaneohe	Thomas Sakamoto	(808) 235-2616

IDAHO
Idaho Falls	William Nielsen	(208) 529-0111
Lewiston	Mel Streeter	(208) 746-2341 ext. 220
Tetonia	William Scott	(208) 456-2233

ILLINOIS
Champaign	Eddie Lane	(217) 356-6996
Chicago	George Sopocko	(312) 545-3622
Hoffman Estates	Chester Dunn	(312) 885-1613
Oak Park	Paul Tan	(312) 848-6327

INDIANA

Evansville	Thomas Peck	(812) 477-4552
		(812) 259-4960
Hanover	Charles Lenhart	(812) 866-4980
Indianapolis	Frank Teskey	(317) 926-1486
Mishawaka	Louis Beehler	(219) 259-4960
Sellersburg	Keith Noe	(812) 246-3301
Vincennes	Arval Donovan	(812) 886-4666
Warsaw	Donald R. Faas	(219) 267-9470

IOWA

Des Moines	Leonard Bowdre	(515) 964-6484
Montpelier	Harold Dugan	(319) 381-2574
Sioux City	Robert L. Hinders	(712) 276-0389 ext. 279

KANSAS

Beloit	Stanley Creitz	(913) 738-2279
Kansas City	Michael Baughman	(913) 334-1000
Liberal	Keith Knos	(316) 624-5908
Wichita	Dr. G.W. Coconis	(316) 686-5611
	John Krier	(316) 722-4041

KENTUCKY

Barbourville	Jack Sellards	(606) 546-5224
Flatwoods	Jerry Middleton	(606) 928-6427
Louisville	Edward Kimmel	(502) 451-3457
Pineville	Edward Ondo	(606) 337-2346

LOUISIANA

Kenner	Don Creech	(504) 467-4644
	Richard C. Rhode	(504) 469-8601
Lafayette	Steven Lumpkin	(318) 233-6941
Metairie	Malcolm McCann, Jr.	(504) 837-7272
Shreveport	James Sorrels	(318) 686-4637
West Monroe	W.H. Hartzfeldt	(318) 396-7431

MAINE

Madawaska	Aurel Daigle	(207) 728-4304
Presque Isle	Gordon Burgess	(207) 769-2461
South Portland	Ted Stackhouse	(207) 799-2238
Topsham	Walt Wheeler	(207) 353-2117

MARYLAND
Baltimore	Earl Tickler	(301) 744-5769
Hyattsville	Carl Miller	(301) 864-5750

MASSACHUSETTS
Cambridge	Frank Serra	(617) 492-2661
Waltham	Robert Ciuffetti	(617) 890-7711
Woburn	A.A. Bellavia	(617) 935-3838
	Mark Ricciardi	(617) 273-3545

MICHIGAN
Ann Arbor	Bob Bellers	(313) 973-3316
Big Rapids	Willard Rush, Jr.	(616) 796-0461 ext. 4895
Dearborn	Adolph Moskal	(313) 834-7010
Detroit	John J. Spurlin	(313) 496-2691
Livonia	Michael Anderson	(313) 591-3833
Lansing	Walter Reilly	(517) 485-9551
Northern Michigan	Gerald H. Heyn	(906) 227-2926
University		(906) 249-9379
Port Huron	John Borris	(313) 984-3881 ext. 305
Warren	Dr. Joel Goldberg	(313) 445-7455
	Eugene Ranta	(313) 445-7343

MINNESOTA
Duluth	Robert Hendrickson	(218) 722-2801
	Ray Pearson	(218) 724-1798
Eden Prairie	Norm Bixby	(612) 944-2222 ext. 284
Forest Lake	Robert Brandell	(612) 421-1397
Minneapolis	Gene Haugen	(612) 330-2400
	David Lewicki	(612) 370-9400 ext. 299
	James D. Mecklin	(612) 522-8527
Moorhead	Orville Kolstad	(218) 233-6496
Mankato	Loren Johnson	(507) 625-3624
	Doug Laven	(507) 625-3441 ext. 85
St. Hilaire	Ron Konickson	(218) 681-5424
Waseca	Dale Keeney	(507) 835-1252

MISSISSIPPI
Goodman	Robert Arnett	(601) 472-2312
Pontotoc	Jimmie G. Collums	(601) 489-2191

MISSOURI
Bethany	Jim Hunsucker	(816) 425-2201
Chillicothe	Robert Chenoweth	(816) 646-3414
Platte City	James Everett	(816) 464-5505

NEBRASKA
Omaha	Jack Hopson	(402) 556-4018

NEVADA
Las Vegas	Bernard E. Farrow	(702) 295-6855
	F.H. Freeman	(702) 295-1915
	Bill Horton	(702) 438-1828

NEW HAMPSHIRE
Hollis	James Belanger	(603) 465-2422
Portsmouth	Albert Cloutier	(603) 436-9011

NEW JERSEY
Cinnaminson	Joseph Szumowski	(609) 829-8957
Lyndhurst	Ian Simms	(201) 843-8500
Malaga	Cyrus Dinshah	(609) 694-3555
Vincetown	Joseph Carney, III	(609) 859-0409
West Milford	Raymond Unser	(201) 697-2072

NEW MEXICO
Las Cruces	Dr. Ronald Hartman	(505) 522-6533
Truth or Consequences	Robert Whittington	(505) 894-3442

NEW YORK
Albany	Warren Baker	(518) 462-2309
Amityville	Murray Barlowe	(516) 691-8733
Big Flats	Edward J. Flynn	(607) 562-3243
Clarence	Ralph Van Sickle	(716) 759-8198
Dundee	Walter Whitehead	(607) 243-7207
Glen Falls	Hubert West	(518) 793-4491
New York	Larry Stockler	(212) 777-6400
Patchogue	Herbert Grossman	(516) 286-5819
Poughkeepsie	Ronald Palluth	(914) 462-2120
Utica	John Earle	(315) 792-5400
Wading River	Joseph Gatz, II	(516) 929-8109
Yorktown Heights	Joseph Passaretti	(914) 926-4376

NORTH CAROLINA

Arden	Wayne Smith	(704) 225-7671
Boone	William H. Mast	(704) 262-3110
Dobson	Joseph G. Sloop	(919) 386-8121
Fayetteville	Melvin E. Talbert	(919) 483-0801
Gastonia	Keith Law	(704) 864-3269
Greensboro	W.M. Brandon	(919) 852-3976
Jacksonville	Ralph Pollmiller	(919) 455-2828
Spruce Pine	Jerry Cox	(704) 765-7351
	Roy Ray, Jr.	(704) 765-7351
Wentworth	Lanny Logan	(919) 342-4261 ext. 180

NORTH DAKOTA

West Fargo	Lloyd Chale	(218) 236-6277

OHIO

Akron	Larry L. K'Meyer	(216) 923-9959
	J.R. Manchester	(216) 644-8500
Cleveland	Stuart Coffe	(216) 391-9696
	Gary Rathbun	(216) 391-9696
	Ottie Sypolt	(216) 987-4692
Columbus	Donald Sisk	(614) 235-0006
Painesville	Andrew Flock	(216) 696-2626
Stow	Anthony Falcione	(216) 688-2451
Toledo	David Garwacki	(419) 475-9221
Xenia	Eddie Gentry	(513) 372-8781
		(513) 426-0291 Dayton
Youngstown	John Carroll	(216) 757-9873
	Stan Jones	(216) 788-8247

OKLAHOMA

Tulsa	Hubert Wood	(918) 494-4031

OREGON

Klamoth Falls	Charles Uhlig	(503) 884-5512
Medford	Orlan Stone	(503) 772-8266
N. Bend	Henry Fuller	(503) 756-6811
Ontario	Curtis Knight	(503) 889-2159

Pendleton	William Manny	(503) 276-3152
Portland	Larry Broschart	(503) 253-9017
	Vern Hartshorn	(503) 667-7292
	Robert L. Jeffery	(503) 285-7068
	Richard Saunders	(503) 271-2204
Redmond	Cecil Peo	(503) 548-4078
Roseburg	George W. Day	(503) 440-4774
Silverton	Karl Moore	(503) 873-2946

PENNSYLVANIA

Allentown	LeRoy Leibensperger	(215) 395-5891
Altoona	H.E. Montgomery	(814) 946-2705
Bristol	Henry M. Wachowicz	(215) 824-3200
Broomall	Michael Brooks	(215) 352-5586
Mountaintop	Gene Hedgepeth	(717) 868-6566
Philadelphia	Joseph Bradley	(215) 561-8046
	Thomas Matthews	(215) 637-5368
Pittsburgh	Dr. M. Thanna Khalil	(412) 391-4100
Wilkes-Barre	Carl Eddy	
York	Luther W. Stone	(717) 757-1100

RHODE ISLAND

Pawtucket	Thomas Plant	(401) 723-3500
		(401) 725-8719

SOUTH CAROLINA

Beaufort	Robert J. Williams	(803) 846-7209
		(803) 524-4919
Darlington	Bob Scott	(803) 393-2616
Greenville	Harry McCall	(803) 294-0280

SOUTH DAKOTA

Pierre	David Merrill	(605) 224-4701
Rapid City	Keith Pruitt	(605) 343-0434
Sioux Falls	Barbara Streemke	(605) 339-0100
Watertown	Donald Haag	(605) 886-8932

TENNESSEE

Bartlett	Charley F. Ray	(901) 382-9338
Chattanooga	James R. Frith	(615) 867-7966
		(615) 899-9545
Knoxville	Ronald Rackley	(615) 688-9422
	Ron Servies	(615) 523-6545
	William Warren	(615) 546-1121
Sparta	William Davis	(615) 738-8423

TEXAS

Austin	Art Ruppert	(512) 452-9403
Corpus Christi	Preston Wallace	(512) 881-6402
Dallas	David Johnson	(214) 263-2613
	Frank Kasal	(214) 233-5757
	Charles W. Kelley	(214) 651-1731
	F. Gilbert Smith	(214) 824-2047
El Paso	Reiner Junge	(915) 566-9621
	Donald Turner	(915) 755-4888
Fort Worth	ISCET	(817) 921-9101
	Robert Chewning	(817) 246-9623
	Robert M. Griffin	(817) 534-4861 ext. 373
Georgetown	Art Ruppert	(512) 863-9157
Houston	Clarence G. Bennett	(713) 656-3318
	Joseph H. Benoit	(713) 656-2775
Irving	Thomas Underwood	(214) 258-1167
Jacksonville	Edward Lewis	(214) 586-9493
Killeen	Ronald Witcher	(817) 526-1247
Lubbock	J.W. Newsom	(806) 763-8246
San Antonio	Gerald Martin	(512) 496-1134
	George Smith	(512) 828-1306
Sherman	Joseph Hudson	(214) 892-9356
Slaton	Darris Linder	(806) 828-5662
Sweetwater	Mabry Tatom	(915) 235-8441
Temple	James Harris	(817) 773-1439
Tyler	Jim Hensley	(214) 535-1480
Waco	Frank Sosolilk	(817) 799-3611
Webster	Freeman Douglas	(713) 486-1616
Wichita Falls	Gene Shearman	(817) 767-4802

VERMONT
Salisbury   W. Clem Small   (802) 352-9018

VIRGINIA

Ararat	Joseph Sloop	(703) 251-5753
Culpepper	Leonard Cowherd	(703) 825-0476
Fredericksburg	Raymond W. Gill	(703) 274-3947
Hardy	Wilfred Davis	(703) 890-1322
Lynchburg	Jerry Hartley	(804) 239-0321 ext. 311
Martinsville	Phillip Jones	(703) 632-5045
Norfolk	Jim Teeters	(804) 588-5975
Richlands	Daniel Bowling	(703) 964-2555

❖ Appendix I
ETA Exam Administrators for Certification

STATE	CITY	CERTIFICATION ADMINISTRATOR
ALABAMA	Deatsville	Baby L. Hall, PO Box 209 36022
	Montgomery	Harold L. Coomes, John M. Patterson State Technical College 3920 Troy Highway 36116
	Tallassee	Ken Hornsby, 304B John St. 36078
ALASKA	Kotzebue	Edward David Sullivan, PO Box 330 99752
ARIZONA	Sierra Vista	Dr. Goodwin Petersen, CET, 148 Sierra Grande 85635
ARKANSAS	Conway	Gayle Glover, CET, R.R. #1, Box 323 A 72032

STATE	CITY	CERTIFICATION ADMINISTRATOR
CALIFORNIA	Apple Valley	Daniel C. Harley, United Western Electronics, 21775 Hwy. 18, Bldg. B4 92307
	Campbell	Cliff Hansen, Westgate TV—Stereo Serv. Center, Hamilton Square, 841 West Hamilton Ave. 95008
	Cucamonga	Phil Lund, CET, 9028 Archibald 91730
	Hayward	Earven F. Horton, 21845 Mission Bldg. 94541
	Lancaster	Walter Nelson, CET, 859½ W. Ave. I 93534
	Pleasant Hill	Charles Petersen, American Monitor Corp., 70 Doray Drive 94523
	San Jose	John Feagin, CET, 56 Boston Ave., #3 95128
COLORADO	Arvada	David Reynolds, Lake Arbor, 8221 Newland Circle 80003
	Greeley	Fred Bantin, Technical Division, AIMS Community College, P.O. Box 69 80632
	Nucla	Alan B. Greager, CET, P.O. Box 16 81424
CONNECTICUT	Hamden	Pat Carangelo, 584 Gilbert Ave. 06514
	New Haven	Albert Marcarelli, Connecticut School of Electronics, 586 Boulevard 06519
FLORIDA	Bushnell	Arthur J. Crawford, Electronics Institute Florida Dept. of Corrections, Sumter Correction Institution, PO Box 667 33513
	Clearwater	Dennis L. Miles, Pinellas Vocational Technical Institute, 6100—154th Ave. 33520
	Fort Lauderdale	Edward Guary, CET 1110 N.E. 4th Ave., 33304
	Gainesville	Charles Couch, CET 1305 N.E. 7th Ter. 32601
	Jacksonville	John Bracher, CET, 2268 Atlantic Boulevard 32207
	Largo	Patrick L. Reilly, 401 Rosery Rd. NE #615 33540
	Miami	Theodore Riener, CET, 9840 S.W. 23rd Terrace 33165
	Mt. Dora	Jim Clower, 1515 Morningside Dr. 32757

STATE	CITY	CERTIFICATION ADMINISTRATOR
	Patrick AFB	B.H. Flowers, RCA International Service Co. POB 4308 Antigua 32925
	Temple Terrace	James Russell, 6120 Whiteway Dr. 33617
GEORGIA	Augusta	Robert Jewett, 1803 McDowell St. 30904
	Atlanta	Harrie Buswell, CET, Chemistry Dept., Georgia Tech 30332
	Brunswick	Dorman McDonald, CET, 3411 Norwich St. 31520
	Clarkesville	Roy Chastain, NGA Tech-voc School, Electronics Dept., Burton Road 30523
	Perry	Hershall Lawhorn, 1013 Main St. 31069
IDAHO	Coeur D'Alene	Jack Bennett, CET, 1602 North 4th 83814
	Fruitland	Curtis E. Knight, PO Box 463 83619
	Meridian	Paul R. Jansson, 600 East Columbia 83642
	Middleton	Vernon Honey, P.O. Box 326 83644
	Twin Falls	Mel Quale, CET, 1730 Kimberly Rd. 83301
	Twin Falls	Francis True, CET, 1516 Addison Ave., E. 83301
ILLINOIS	Champaign	Eddie Lane, CET, 1501 Honeysuckle 68120
	Chicago	Don Thorne, 3733 W. Eddy St. 63618
	Moline	Lester Knapp, 3431 60th., Apt. 3C 61265
	Troy	Gene Fayollat, Tri-county TV, 112 E. Market 62294
	Washington	Robert J. Petzing, AVC Electronics, 1411 Washington Road 61571
INDIANA	Bloomington	Robert E. Drake, 2320 Evergreen 47401
	Connersville	Leonard Smith, Electronic Department, Connersville Area Vocational School, 1000 Ranch Road 47331
	Evansville	Thomas Peck, CET, Brewsters TV, 1401 Convert Ave. 47714
	Greencastle	Dick Glass, Sr., CET, R.R.3, Box 564 46135
	Greenfield	Grover Harvey, R.R. #8, Box 11 46140
	Griffith	Elbert Powers, CET, 139 N. Griffith Blvd. 46319
	Hanover	Charles William Lenhart, R.R. #1, Box 61 47243

STATE	CITY	CERTIFICATION ADMINISTRATOR
	Hartford City	James A. Smith, CET, 409 W. Commercial St. 47348
	Indianapolis	Leon Howland, CET, 4624 E. 10th St. 46201
	Indianapolis	Dick Schultz, c/o III Technical Institute, 1720 East 38th 46218
	Jeffersonville	O.C. Brown, 413 Mockingbird Dr. 41730
	Terre Haute	Larry Lowey, Indiana Voc. Tec. College, 7377 S. Dixie Bee Road 47802
	Versailles	Charles Workman, SIVS, PO Box 156 47042
IOWA	Ames	Ron Crow, CET, ERI Electronics Services, 131K Coover Hall ISU 50011
	Ankeny	John Arbuckle, Electronics Instructor, Des Moines Area Community College, 2006 Ankeny Blvd. 50021
	Davenport	Del Menke, CET 3526 Fair Ave. 52806
	Marshaltown	Donald Anker, CET, Fisher Controls 50158
LOUISANA	New Orleans	Gary M. Raymond, CET, 6330 Pratt Drive 70211
MARYLAND	Baltimore	Earl Tickler, CET, 1004 Hallimont 21228
	Baltimore	David Windisch, 6218 Mossway 21212
	Linthicum	Jesse B. Leach, Jr., CET, 231 Hammonds Ferry Road 21090
MASSACHUSETTS	Ashland	Ted Soboscienski, CET, 26 Carriage House Path 01721
MICHIGAN	Ann Arbor	Bob Bellers, CET, 4800 E. Huron River Dr. 48106
	Flint	Bruce Miller, CET, Mott College 48503
	Lansing	Walter Reilly, CET, 2301 Groesbeck Avenue 48912
	Mason	Jesse Ramey, 932 Eugenia 48854
MINNESOTA	Austin	Vincent Lynch, 1900 8th Avenue, N.W. 55912
	Brooklyn Park	Robert Brandell, Hannepin Technical Center, North, 9000 North 77th Ave. 55429

STATE	CITY	CERTIFICATION ADMINISTRATOR
	Duluth	Ray S. Pearson, CET, 329 W. Fairbault St. 55803
	Faribault	John Baldwin, 1017 SW 8th Ave. 55021
MISSISSIPPI	Goodman	Robert Arnett, CET, Holmes Junior College 39079
MISSOURI	Gladston	George De Lisle, MaCET, 603 NE 73rd Terrace 64118
	St. Louis	James R. Long, CET 10409 Coburg Lands Drive 63137
	St. Louis	Vincent J. Lutz, CET, 1546 Sells Ave. 63147
	St. Louis	Robert Paynter, Bailey Technical School, 3750 Lindell Bldg. 63108
MONTANA	Missoula	F.W. Boisvert, 120 Mary Ave. 59801
NEBRASKA	Doniphan	George Savage, CET, Box 39 68832
	Hastings	Eugene L. Young, CET, Box 1024 68901
	Lincoln	Olin Boone, 6927 Holdrege 68505
	North Platte	Gordon P. Koch, CET, Mid-Plains Voc. Tech. School, I 80 and Highway 83 69101
	Omaha	John D. Snyder, CET, 3221 S. 45th St. 68106
NEW HAMPSHIRE	Portsmouth	Albert Cloutier, CET, 369 Islington St. 03801
NEW YORK	Fishkill	Peter Moranski, CET, 35 Revere Road 12524
NORTH CAROLINA	Albermarle	Ricky W. Bean, CET, 700 South Fifth St. EXT 28001
	Charlotte	Thomas L. Ruth, 1511 Pierson Dr. 28205
	Granite Quarry	Ted Morton, CET, P.O. Box 306 28072
	Hickory	Ralph Hartley, CET, R.R. #10, Box 901 28601
	Statesville	Wilson Allsbrook, CET, Mitchell Comm. School, West Broad St. 28677
	Statesville	Dorman (Bud) Tatum, 944 David Ave. 28677
OHIO	Canton	Hal Frutschy, CET, 921 29th, NE 44714
OKLAHOMA	Stillwater	Jerry Don Harris, CET, 1002 So. Main 74074

STATE	CITY	CERTIFICATION ADMINISTRATOR
	Oklahoma City	Phillip B. Morris, 3501 N.W. 65th 73116
	Oklahoma City	Cliff Sheffield, CET, c/o Jack Broadford TV & Appliance, 3101 N. May 73112
OREGON	Burns	F.L. Goddard, PO Box 713 97720
	Lakeside	Wm. D.C. Burnette, CET, PO Box 451 97499
	Salem	Al Stratton, Television & Radio Licensing, 4th Floor, Labor & Industry Bldg. 97310
PENNSYLVANIA	Dunmore	Ronald Lettieri, MaCET, 433 East Drinker St. 18512
	Lafayette Hill	John Guinan, CET, 2178 Joshua Road 19444
	St. Lawrence	Stanley Golowski, CET, 3614 Orchard Court 19606
	Scranton	William Hessmiller, CET, 810½ Grandview St. 18509
SOUTH CAROLINA	Columbia	William J. Rivers, CET, 1451 Bonner Ave. 29204
TENNESSEE	Gibson	Donald Bartholomew, CET, PO Box 646 38338
	Shelbyville	Larry Haggard, CET, PO Box 512 37160
TEXAS	Beaumont	Richard F. Arsenault, 3190 Eastex Freeway #102 77703
	College Station	Basil Collins, CET, Inst of Elec. Science, Texas A & M University, F. E. Drawer K 77843
	Houston	D. C. Larson, CET, 1308 Aldrich 77055
	Houston	David Van Winkle, 2700 WW Thorn Drive 77073
	Killeen	Ron Wichter, Central Texas College, Highway 190 West 76541
	Slaton	Darris D. Linder, CET, 340 South 12th St. 79364
	Spring Branch	Roger Book, CET, Star Route 1, Box 450 78070
VIRGINIA	Colonial Heights	Curtis E. Anderson, Jr., CET, 4814 Conduit Road 23834
	Emporia	Carl B. Rae, Jr., CET, PO Box 791 23847

STATE	CITY	CERTIFICATION ADMINISTRATOR
	Grafton	John McPherson, CET, PO Box 1347 23692
	Hampton	Walter R. Cooke, CET, 957 North King St. 23669
	Lynchburg	Jerry Hartley, CET, 220 Chesterfield Road 24502
	Martinsville	Phillip M. Jones, CET, 616 Liberty St. 24112
	Norfolk	James Tetters, 2937 East Malden Ave. 23518
	Wyers Cave	Leon Smith, Blue Ridge Community College, PO Box 80 24486
WASHINGTON	Colville	Herber Spille, CET, 33 North Elm 99114
	East Wenatchee	Roberd Eley, 125 26th, N.W. 98801
	Hoquiam	Alfred L Izatt, 2315 Queens 98550
	Seattle	Dean Thompson, 3850 N.E. 95th 98115
WISCONSIN	Appleton	Larry Mattioli, CET, Fox Valley Technical Institute, 1825 North Bluemound Drive 54913
	LaCrosse	R.F. Thomas, CET, Thomas Electronics, 1202 West Ave. South 54601
	Madison	Duane Busby, CET, 2027 Sherman Ave. 53104
	Sheboygan	Robert H. White, CET, 520 Greendale Rd, 53081
	Wausau	Jacob Klein, CET, 3419 Polzer Drive 54401
WYOMING	Casper	Martin E. Engle, 5018 Alcova South, Box 11 82601
CANADA	Montreal Quebec	Heikki Thoen, 7475 Sherbrooke St. W. H48 IS4
	Victoria, BC	Gordon H. Bjornson, CET, 1564 Agnew Ave. V8N 5M5
	Waterloo Ontario	Ray Pierce, 1-1202-300 Regina St. N., N21 3B8

❖ Appendix J
Society of
Broadcast Engineers

Local Chapters and Chairmen

Birmingham, Alabama #68
Paul D. Gross
WVTM-TV
PO Box 10502
Birmingham, AL 35202

Anchorage, Alaska #89
Frank Mengel
SBE Chapter 89
PO Box 104338
Anchorage, AL 99510

Phoenix, Arizona #9
William Strube
KPHO-TV
4016 N. Black Canyon Hwy.
Phoenix, AZ 85017

Tuscon #32
Richard Heatley
KZAZ-TV
2445 N. Tucson Blvd.
Tucson, AZ 85716

Eureka, California #71
Donald R. King
KIEM-TV
5650 S. Broadway
Eureka, CA 95501

Fresno #66
Hal Torosian
SBE Chapter #66
PO Box 12965
Fresno, CA 93779

Los Angeles #47
Tim Schultz
KMEX-TV
5420 Melrose Avenue
Hollywood, CA 90038

Orange County #77
Don Beem
520 Brenthaven
Anaheim, CA 92801

Sacramento #43
Bob Hess
KOVR-TV
1216 Arden Way
Sacramento, CA 85815

San Diego #36
Tom Wimberly
KCST-TV
8330 Engineer Road
San Diego, CA 92111

San Francisco Bay #40
Art Lebermann
KRE/KBLX
601 Ashby Ave.
Berkeley, CA 94710

Denver, Colorado #48
Gerry Westerberg
2141 E. 83rd Place
Denver, CO 80229

Fort Collins #50
Estel Haning
EMS-Michener Bldg. L 171
Univ. of N. Colorado
Greeley, CO 80639

Grand Junction #81
Richard Cron
407½ Bristol Court
Grand Junction, CO 81501

Connecticut Valley #14
John Reno
WFSB-TV
3 Constitution Plaza
Hartford, CT 06115

Stamford #92
Steven Mendel
Group W Satellite Comm.
PO Box 10210
Stamford, CT 06904

District of Columbia #37
Dick Cassidy
National Public Radio
2025 "M" St. NW
Washington, D.C. 20036

Central Florida #42
Bob Diehl
WCPX-TV
PO Box 66000
Orlando, CL 32853

North Central Florida #95
Gres Whitsett
WUFT-TV
1020 Weimer Hall, U of Florida
Gainsville, FL 32611

Southwest #90
Jim Eblen
WKZY
3440 Marina Towne Lane
N. Fort Myers, FL 33903

Jacksonville #7
Steve Flanagan
WJXT-TV
1851 Southampton Road
Jacksonville, FL 33207

Palm Beach #88
George Danner
WPEC-TV
Fairfield Drive
West Palm Beach, FA 33407

South Florida #53
Henry Seiden
WPLG-TV
3900 Biscayne Blvd.
Miami, FL 33137

Tampa Bay Area #39
Ralph Beaver
WRBQ Radio
5510 Gray St.
Tampa, FL 33609

Atlanta, Georgia #5
Victor Jester
WSB Radio
1601 W. Peachtree St. N
Atlanta, GA 30309

Honolulu, Hawaii #63
Robert N. Palitz
Caughill-Palitz
1750 Kalakaua Ave. #3-120
Honolulu, Hawaii 96826

Chicago, Illinois #26
Steve Gordoni
Chacago SBE Chapter
121 W. Wacker Dr.
Chicago, IL 60601 ·

Central Illinois #49
Bill Beatty
TeleVisual, Inc.
1287 Wabash Avenue
Springfield, IL 62704

John Heidermann
Video Midwest, Inc.
2906 Brady St.
Davenport, IA 52803

Indianapolis, Indiana #25
Doug Garlinger
WHMB
10511 Greenfield Ave.
Noblesville, IN 46060

South Bend Michigan #30
Russ Summerville
WNDU
PO Box 1616
South Bend, IN 46634

Kansas #3
Gary Krohe
KLDH-TV
PO Box 2229
Topeka, KS 66601

High Plains #94
Kenneth A. Sell
KSNG TV
South Star Route
Garden City, KS 67846

Kentucky #35
Tom Landers
WKPC-TV
PO Box 37380
Louisville, KY 40233

Baltimore, Maryland #46
Michael A. Fast
WPOC
711 W. 40th St.
Baltimorre, MD 21211

Boston, Massachusetts #11
Doug Sorensen
Datatron, Inc.
48 Royal Crest Dr. #10
North Andover, MA 01845

Central Michigan #91
Larry Estlack
Central Michigan SBE Chapter
PO Box 22053
Lansing, MI 48909

Southeast Michigan #82
Paul Grzebik
WQRS-FM
500 Temple Ave.
Detroit, MI 48201

Minneapolis/St. Paul Minnesota #17
Peter L. Thorpe
B-108 FM
PO Box 1470
Anoka, MN 55303

Jackson, Mississippi
Mike Hughes
WJDX
PO Box 2171
Jackson, MS 39205

Kansas City, Missouri #59
Jack Braton, Vice/Chrm
WDAF
3030 Summit
Kansas City, MO 64108

St. Louis Area #55
Jim Jackson
KWK Radio
2360 Hampton Ave.
St. Louis, MO 63139

Springfield
Bill Martin
W.B. Martin & Assoc.
RT 4, Box 52
El Dorado Springs, MO 64744

Holdrege, Nebraska #87
Jerry Fuehrer
KHGI-TV
PO Box 220
Kearney, NE 68847

Midland #74
David Messing
KNCY-AM-FM
PO Box 278
Nebraska City, NE 68410

Alibuquerque, New Mexico #34
Lothar Merker
6840 Frantz NE
Albuquerque, NM 87109

Binghamton, New York #1
Douglas S. Colborn
724 W. Gray St.
Elmira, NY 14905

Central New York #22
John Soergel
25 Cotty Drive
East Syracuse, NY 13057

New York City #15
Lyn Synder
Box 182
Floral Park, NY 11001

Northeast New York #58
Charles Zarriello
17 Briarwood Terrace
Albany, NY 12203

Rochester #57
Jeff Baker
WBBF-WMJQ
850 Midtown Tower
Rochester, NY 14604

Charlotte, North Carolina #45
W.C. Groves, Jr.
1219 S. Oakwood Ave.
Gastonia, N.C. 28052

Raleigh-Durham #93
Gary Liebisch
106 Hollow Oak St.
Cary, NC 27511

Winston-Salem #84
Edward J. Kasovic, Jr.
WJTM-TV
3500 Myer Lee Dr.
Winston-Salem, N.C. 27101

Athens, Ohio
David W. Palmer
WATH/WXTQ
300 N. Columbus Rd.
Athens, OH 45701

Central Ohio #52
David W. White
2888 Larrimer Ave.
Worthington, OH 43085

Northeast Ohio #70
Soc. of Best. Engs.
Chpt. #70
PO Box 91272
Cleveland, OH 44101

Southwestern Ohio #33
Don Pagan
3576 Haney Road
Dayton, OH 45416

Central Western Oklahoma #85
Greg Miller
KGMC-TV
PO Box 14587
Oklahoma City, OK 73113

Tulsa #56
William R. Schock
24 Trailridge Rd.
Sabulpa, OK 74066

Portland Oregon
Larry Wilson
KXL Radio
1415 S.E. Ankeny
Portland, OR 97214

Central Pennsylvania #41
Gary Reed
42 Vine St.
Highspire, PA 17034

Northeastern Pennsylvania #2
Ronald Schacht
WNAK Radio
84 S. Prospect St.
Nanticoke, PA 18643

Philadelphia #18
Robert B. Hoy
WWDB
3930 Conshohocken Ave.
Philadelphia, PA 19131

Pittsburgh #20
Mark Albright
WPGH-TV
750 Ivory Avenue
Pittsburgh, PA 15214

Charleston, South Carolina
Lowell Knouff
WCSC-TV
PO Box 186
Charleston, S.C. 29402

Greenville Area #86
Jobie Sprinkle
7A Ector St.
Asheville, N.C. 28806

Nashville, Tennessee
George McClintock
WNQM
3314 West End Ave.
Nashville, TN 37203

Memphis #61
Pat Lane
WKNO-TV/FM
Box 80,000-MSU
Memphis, TN 38152

Central Texas #79
George Taylor
KBVO-TV
PO Drawer 2728
Austin, TX 78768

Corpus Christi #29
John Ross
4750 S. Padre Island Dr.
Corpus Christi, TX 78411

El Paso #38
Carl Bahner
KTSM
801 N. Oregon
El Paso, TX 79902

North Texas #67
Jack Sellmeyer
SBE Chapter 67
PO Box 300532
Arlington, TX 76010

South Texas #69
Frank Fleanor
4319 Casa Manana
San Antonio, TX 78233

Blue Ridge, Virginia #78
Old Time Gospel Hour TV
Milton S. Ridgeway
Tech. Operations Mgr.
701 Thomas Rd.
Lynchburg, VA 24514

John Francioni
WRVA Radio
P.O. Box 1516
Richmond, VA

Tidewater #54
Gene Gildow
WTKR-TV
PO Box 2456
Norfolk, VA 23501

Seattle, Washington #16
Allen Hartle
KZOK Suite #304
300 West Mercer
Seattle, WA 98119

Spokane #21
Gordon Canaday
KMBI Radio
PO Box 8024
Spokane, WA 99203

Tri Cities #51
Felipe E. Olivera
KVEW-TV
601 N. Edison
Kennewick, WA 99336

Southern West Virginia
Pete Stark
WPBY-TV
Marshall University
Huntington, WV 25701

Fox Valley, Wisconsin #80
Gary Mach
UWBG TeleProduction Center
UW-Green Bay
Green Bay, WI 54302

Madison #24
Doug C. McDonnell
2830 Waubesa Ave.
Madison, WI 53711

Milwaukee #28
Terrence Baun
WEZW
735 W. Wisconsin Ave.
Milwaukee, WI 53233

Kitchener, Ontario Canada
Paul Firminger
CHYM/CKGL-FM
305 King St. West
Kitchener, Ontario N2G 4F4
CANADA

Index